THEA Math Practice Book

Complete Content Review Plus 2 Full-length THEA Math Tests

By
Elise Baniam & Michael Smith

THEA Math Practice Book

Published in the United State of America By

The Math Notion

Email: info@Mathnotion.com

Web: WWW.MathNotion.com

Copyright © 2020 by the Math Notion. All rights reserved. No part of this publication may be reproduced, stored in a retrieval system, or transmitted in any form or by any means, electronic, mechanical, photocopying, recording, scanning, or otherwise, except as permitted under Section 107 or 108 of the 1976 United States Copyright Ac, without permission of the author.

All inquiries should be addressed to the Math Notion.

ISBN: 978-1-63620-187-0

About the Author

Elise Baniam has been a math instructor for over a decade now. She graduated in Mathematics. Since 2006, Elise has devoted his time to both teaching and developing exceptional math learning materials. As a math instructor and test prep expert, Elise has worked with thousands of students. She has used the feedback of her students to develop a unique study program that can be used by students to drastically improve their math score fast and effectively.

– **HiSET Math Workbook**
– **TASC Math Workbook**
– **ASVAB Math Workbook**
– **AFOQT Math Workbook**
–**many Math Education Workbooks**
– **and some Mathematics books ...**

As an experienced Math teacher, Mrs. Baniam employs a variety of formats to help students achieve their goals: she teaches students in large groups, and she provides training materials and textbooks through her website and through Amazon.

You can contact Elise via email at:
Elise@Mathnotion.com

THEA Math Practice Book

THEA Math Practice Book is **an excellent investment in your future** and the best solution for students who want to maximize their score and minimize study time. Practice is an essential part of preparing for a test and improving a test taker's chance of success. The best way to practice taking a test is by going through lots of THEA math questions.

High-quality mathematics instruction ensures that students become problem solvers. We believe all students can develop deep conceptual understanding and procedural fluency in mathematics. In doing so, through this math workbook we help our students grapple with real problems, think mathematically, and create solutions.

THEA Math Practice Book allows you to:

- Reinforce your strengths and improve your weaknesses.
- Practice **2500+ realistic** THEA math practice questions
- Exercise math problems in a variety of formats that provide intensive practice.
- Review and study **Two full-length THEA practice tests** with detailed explanations

...and much more!

This Comprehensive THEA Math Practice Book is carefully designed to provide only that **clear and concise information** you need.

Get the Targeted Practice You Need to Excel on the Math Section of the THEA Test!

WWW.MathNotion.com

… So Much More Online!

✓ FREE Math Lessons

✓ More Math Learning Books!

✓ Mathematics Worksheets

✓ Online Math Tutors

For a PDF Version of This Book

Please Visit WWW.MathNotion.com

Contents

Chapter 1: Whole Numbers 11
    Add and Subtract Integers 12
    Multiplication and Division 13
    Absolute Value 14
    Ordering Integers and Numbers 15
    Order of Operations 16
    Factoring 17
    Great Common Factor (GCF) 18
    Least Common Multiple (LCM) 19
    Divisibility Rule 20
    Answer key Chapter 1 21

Chapter 2: Fundamental 25
    Adding Fractions – Unlike Denominator 26
    Subtracting Fractions – Unlike Denominator 27
    Converting Mix Numbers 28
    Converting improper Fractions 29
    Addition Mix Numbers 30
    Subtracting Mix Numbers 31
    Simplify Fractions 32
    Multiplying Fractions 33
    Multiplying Mixed Number 34
    Dividing Fractions 35
    Dividing Mixed Number 36
    Comparing Fractions 37
    Answer key Chapter 2 38

Chapter 3: Decimal 43
    Round Decimals 44
    Decimals Addition 45
    Decimals Subtraction 46

THEA Math Practice Book

Decimals Multiplication 47
Decimal Division 48
Comparing Decimals 49
Convert Fraction to Decimal 50
Convert Decimal to Percent 51
Convert Fraction to Percent 52
Answer key Chapter 3.. 53

Chapter 4: Equations and Inequality 57

Distributive and Simplifying Expressions 58
Factoring Expressions 59
Evaluate One Variable Expressions 60
Evaluate Two Variable Expressions 61
Graphing Linear Equation 62
One Step Equations 63
Two Steps Equations 64
Multi Steps Equations 65
Graphing Linear Inequalities 66
One Step Inequality 67
Two Steps Inequality 68
Multi Steps Inequality 69
Systems of Equations 70
Systems of Equations Word Problems 71
Finding Distance of Two Points 72
Answer key Chapter 4.. 73

Chapter 5: Exponent and Radicals 79

Positive Exponents 80
Negative Exponents 81
Add and subtract Exponents 82
Exponent multiplication 83
Exponent division 84
Scientific Notation 85
Square Roots 86
Simplify Square Roots 87
Answer key Chapter 5.. 88

THEA Math Practice Book

Chapter 6: Ratio, Proportion and Percent 91
Proportions 92
Reduce Ratio 93
Percent 94
Discount, Tax and Tip 95
Percent of Change 96
Simple Interest 97
Answer key Chapter 6 98

Chapter 7: Monomials and Polynomials 100
Adding and Subtracting Monomial 101
Multiplying and Dividing Monomial 102
Binomial Operations 103
Polynomial Operations 104
Squaring a Binomial 105
Factor polynomial 106
Answer key Chapter 7 107

Chapter 8: Functions 109
Relation and Functions 110
Slope form 111
Slope and Y-Intercept 111
Slope and One Point 112
Slope of Two Points 113
Equation of Parallel and Perpendicular lines 114
Quadratic Equations - Square Roots Law 115
Quadratic Equations - Factoring 116
Quadratic Equations - Completing the Square 117
Quadratic Equations - Quadratic Formula 118
Arithmetic Sequences 119
Geometric Sequences 120
Answer key Chapter 8 121

Chapter 9: Geometry 125
Area and Perimeter of Square 126
Area and Perimeter of Rectangle 127
Area and Perimeter of Triangle 128

Area and Perimeter of Trapezoid 129
Area and Perimeter of Parallelogram 130
Circumference and Area of Circle 131
Perimeter of Polygon 132
Volume of Cubes 133
Volume of Rectangle Prism 134
Volume of Cylinder 135
Volume of Spheres 136
Volume of Pyramid and Cone 137
Surface Area Cubes 138
Surface Area Rectangle Prism 139
Surface Area Cylinder 140
Answer key Chapter 9 141

Chapter 10: Statistics and probability 143
Mean, Median, Mode, and Range of the Given Data 144
Box and Whisker Plot 145
Bar Graph 146
Histogram 147
Dot plots 148
Scatter Plots 149
Stem–And–Leaf Plot 150
Pie Graph 151
Probability 152
Answer key Chapter 10 153

THEA Test Review 157
THEA Practice Test 1 161
THEA Practice Test 2 179

Answers and Explanations 197
Answer Key 199
THEA Practice Test 1 201
THEA Practice Test 2 209

Chapter 1:
Whole Numbers

Add and Subtract Integers

Find the sum or difference.

1) $(+152) + (-98) =$

2) $(+74) + (-42) =$

3) $188 - 85 =$

4) $(-214) + 157 =$

5) $(-72) + 425 =$

6) $182 + (-265) =$

7) $(-15) + 38 =$

8) $415 - 310 =$

9) $(-18) - (-77) =$

10) $(-88) + (-57) =$

11) $(-124) - 304 =$

12) $1{,}420 - (-257) =$

13) $4 + (-15) + (-35) + (-15) =$

14) $(-17) + (-24) + 42 + 12 =$

15) $(-5) - 7 + 38 - 21 =$

16) $8 + (-19) + (-29 - 21) =$

17) $(+45) + (+28) + (-57) =$

18) $(-42) + (-32) =$

19) $-14 - 18 - 9 - 31 =$

20) $9 + (-28) =$

21) $134 - 90 - 53 - (-42) =$

22) $(+37) - (-9) =$

23) $(+7) - (+11) - (-19) =$

24) $(+27) - (+9) - (-32) =$

Multiplication and Division

Calculate.

1) $210 \times 9 =$

2) $160 \times 40 =$

3) $(-6) \times 8 \times (-5) =$

4) $-5 \times (-7) \times (-7) =$

5) $13 \times (-13) =$

6) $40 \times (-8) =$

7) $8 \times (-2) \times 6 =$

8) $(-400) \times (-30) =$

9) $(-20) \times (-20) \times 3 =$

10) $125 \times 6 =$

11) $142 \times 50 =$

12) $364 \div 14 =$

13) $(-4{,}125) \div 5 =$

14) $(-28) \div (-7) =$

15) $288 \div (-18) =$

16) $3{,}500 \div 28 =$

17) $(-126) \div 3 =$

18) $4{,}128 \div 4 =$

19) $1{,}260 \div (-35) =$

20) $3{,}360 \div 4 =$

21) $(-54) \div 2 =$

22) $(-2{,}000) \div (-20) =$

23) $0 \div 870 =$

24) $(-1{,}020) \div 6 =$

25) $5{,}868 \div 652 =$

26) $(-2{,}520) \div 4 =$

27) $10{,}902 \div 3 =$

28) $(-60) \div (-5) =$

Absolute Value

Simplify each equation below.

1) $|-40| =$

2) $-20 + |-40| + 38 =$

3) $|-58| - |-32| + 16 =$

4) $|-8 + 7 - 4| + |5 + 5| =$

5) $3|3 - 9| + 18 =$

6) $|-8| + |-25| =$

7) $|-42 + 18| + 12 - 5 =$

8) $|-16| - |-28| - 7 =$

9) $|-35| - |-14| + 9 =$

10) $|24| - 22 + |-15| =$

11) $\frac{3|4-8|}{4} =$

12) $|-26 + 11| =$

13) $|-24| \times |3| + 4 =$

14) $|-4| + |-28| + 6 - 2 =$

15) $|-16| + |-18| - 19 =$

16) $14 + |-28 + 12| + |-15| =$

17) $26 - |-56| + 20 =$

18) $\frac{|147|}{|7|} + 9 =$

19) $|-6 + 10| + |38 - 18| + 2 =$

20) $|-30 + 18| + |-15| + 10 =$

21) $\frac{|-54|}{6} \times |-9| =$

22) $\frac{6|3 \times 5|}{6} \times \frac{|-12|}{5} =$

23) $\frac{|3 \times 5|}{15} \times 8 =$

24) $|-20 + 4| \times \frac{|-2 \times 6|}{8} =$

25) $|-150 + 10| - 6 + 6 =$

26) $|-80 + 60| - 20 =$

Ordering Integers and Numbers

Order each set of integers from least to greatest.

1) $6, -9, -4, -3, 2$ ___, ___, ___, ___, ___, ___

2) $-4, -18, 5, 14, 11$ ___, ___, ___, ___, ___, ___

3) $29, -28, -16, 27, -21$ ___, ___, ___, ___, ___, ___

4) $-15, -45, 25, -17, 38$ ___, ___, ___, ___, ___, ___

5) $37, -42, 32, -45, 18$ ___, ___, ___, ___, ___, ___

6) $75, 38, -59, 85, -24$ ___, ___, ___, ___, ___, ___

Order each set of integers from greatest to least.

7) $16, 19, -11, -15, -7$ ___, ___, ___, ___, ___, ___

8) $22, 46, -54, -36, 61$ ___, ___, ___, ___, ___, ___

9) $55, -46, -19, 37, -17$ ___, ___, ___, ___, ___, ___

10) $37, 95, -46, -22, 87$ ___, ___, ___, ___, ___, ___

11) $-9, 79, -65, -78, 84$ ___, ___, ___, ___, ___, ___

12) $-70, -35, -50, 17, 39$ ___, ___, ___, ___, ___, ___

Order of Operations

Evaluate each expression.

1) $6 + (2 \times 5) =$

2) $15 - (4 \times 2) =$

3) $(14 \times 5) + 15 =$

4) $(18 - 3) - (4 \times 5) =$

5) $32 + (18 \div 3) =$

6) $(18 \times 4) \div 6 =$

7) $(63 \div 7) \times (-3) =$

8) $(7 \times 8) + (34 - 18) =$

9) $80 + (3 \times 3) + 5 =$

10) $(20 \times 8) \div (4 + 4) =$

11) $(-7) + (12 \times 5) + 11 =$

12) $(5 \times 9) - (45 \div 5) =$

13) $(7 \times 6 \div 2) - (17 + 13) =$

14) $(14 + 6 - 16) \times 8 - 12 =$

15) $(36 - 18 + 30) \times (96 \div 8) =$

16) $24 + \left(14 - (36 \div 6)\right) =$

17) $(7 + 10 - 4 - 9) + (24 \div 3) =$

18) $(90 - 15) + (16 - 18 + 8) =$

19) $(20 \times 3) + (16 \times 4) - 10 =$

20) $15 + 12 - (26 \times 5) + 15 =$

Factoring

Factor, write prime if prime.

1) 16

2) 82

3) 28

4) 46

5) 62

6) 65

7) 38

8) 10

9) 54

10) 85

11) 45

12) 90

13) 81

14) 55

15) 92

16) 114

17) 86

18) 70

19) 105

20) 80

21) 95

22) 36

23) 84

24) 110

25) 34

26) 98

27) 63

28) 106

Great Common Factor (GCF)

Find the GCF of the numbers.

1) 6, 15

2) 46, 27

3) 48, 58

4) 20, 25

5) 16, 36

6) 32, 42

7) 60, 25

8) 90, 35

9) 72, 9

10) 45, 54

11) 88, 54

12) 35, 70

13) 70, 20

14) 32, 82

15) 48, 96

16) 30, 85

17) 16, 24

18) 80, 100, 40

19) 81, 112

20) 56, 88

21) 10, 5, 25

22) 8, 18, 24

23) 15, 45, 60

24) 51, 33

Least Common Multiple (LCM)

Find the LCM of each.

1) 8, 10

2) 32, 16

3) 5, 10, 15

4) 14, 21

5) 20, 4, 30

6) 25, 5

7) 12, 60, 24

8) 5, 6

9) 13, 26, 54

10) 28, 35

11) 27, 54

12) 110, 22

13) 30, 15, 60

14) 18, 63

15) 40, 8, 5

16) 81, 18

17) 38, 19

18) 22, 44

19) 25, 60

20) 16, 48

21) 17, 10

22) 8, 28

23) 35, 70

24) 21, 6

Divisibility Rule

Apply the divisibility rules to find the factors of each number.

1) 15	2, 3, 4, 5, 6, 9, 10		13) 34	2, 3, 4, 5, 6, 9, 10
2) 124	2, 3, 4, 5, 6, 9, 10		14) 385	2, 3, 4, 5, 6, 9, 10
3) 352	2, 3, 4, 5, 6, 9, 10		15) 915	2, 3, 4, 5, 6, 9, 10
4) 94	2, 3, 4, 5, 6, 9, 10		16) 157	2, 3, 4, 5, 6, 9, 10
5) 241	2, 3, 4, 5, 6, 9, 10		17) 540	2, 3, 4, 5, 6, 9, 10
6) 455	2, 3, 4, 5, 6, 9, 10		18) 340	2, 3, 4, 5, 6, 9, 10
7) 65	2, 3, 4, 5, 6, 9, 10		19) 480	2, 3, 4, 5, 6, 9, 10
8) 320	2, 3, 4, 5, 6, 9, 10		20) 3,750	2, 3, 4, 5, 6, 9, 10
9) 1,134	2, 3, 4, 5, 6, 9, 10		21) 660	2, 3, 4, 5, 6, 9, 10
10) 68	2, 3, 4, 5, 6, 9, 10		22) 286	2, 3, 4, 5, 6, 9, 10
11) 754	2, 3, 4, 5, 6, 9, 10		23) 158	2, 3, 4, 5, 6, 9, 10
12) 148	2, 3, 4, 5, 6, 9, 10		24) 456	2, 3, 4, 5, 6, 9, 10

Answer key Chapter 1

Add and Subtract Integers

1) 54
2) 32
3) 103
4) −57
5) 353
6) −83
7) 23
8) 105
9) 59
10) −145
11) −428
12) 1,677
13) −61
14) 13
15)
16) −61
17) 16
18)
19)
20)
21)
22) 46
23) 15
24) 50

Multiplication and Division

1) 1,890
2) 6,400
3) 240
4) −245
5) −169
6) −320
7) −96
8) 12,000
9) 1,200
10) 750
11) 7,100
12) 26
13) −825
14) 4
15) −16
16) 125
17) −42
18) 1,032
19) −36
20) 840
21) −27
22) 100
23) 0
24) −170
25) 9
26) −630
27) 3,634
28) 12

Absolute Value

1) 40
2) 58
3) 42
4) 15
5) 36
6) 33
7) 31
8) −19
9) 30
10) 17
11) 3
12) 15
13) 76
14) 36
15) 15
16) 45
17) −10
18) 30
19) 26
20) 37
21) 81

THEA Math Practice Book

22) 36 24) 24 26) 0
23) 8 25) 140

Ordering Integers and Numbers

1) -9, −4, −3, 2, 6
2) −18, −4, 5, 11, 14
3) −28, −21, −16, 27, 29
4) −45, −17, −15, 25, 38
5) −45, −42, 18, 32, 37
6) −59, −24, 38, 75, 85

7) 19, 16, −7, −11, −15
8) 61, 46, 22, −36, −54
9) 55, 37, −17, −19, −46
10) 95, 87, 37, −22, −46
11) 84, 79, −9, −65, −78
12) 39, 17, −35, −50, −70

Order of Operations

1) 16
2) 7
3) 85
4) −5
5) 38
6) 12
7) −27
8) 72
9) 94
10) 20
11) 64
12) 36
13) −9
14) 20
15) 576
16) 32
17) 12
18) 81
19) 114
20) −88

Factoring

1) 1, 2, 4, 8, 16
2) 1, 2, 41, 82
3) 1, 2, 4, 7, 14, 28
4) 1, 2, 23, 46
5) 1, 2, 31, 62
6) 1, 5, 13, 65
7) 1, 2, 19, 38
8) 1, 2, 5, 10
9) 1, 2, 3, 6, 9, 18, 27, 54
10) 1, 5, 17, 85
11) 1, 3, 5, 9, 15, 45
12) 1, 2, 3, 5, 6, 9, 10, 15, 18, 30, 45, 90
13) 1, 3, 9, 27, 81
14) 1, 5, 11, 55
15) 1, 2, 4, 23, 46, 92
16) 1, 2, 3, 6, 19, 38, 57, 114
17) 1, 2, 43, 86
18) 1, 2, 5, 7, 10, 14, 35, 70
19) 1, 3, 5, 7, 15, 21, 35, 105
20) 1, 2, 4, 5, 8, 10, 16, 20, 40, 80
21) 1, 5, 19, 95
22) 1, 2, 3, 4, 6, 9, 12, 18, 36
23) 1, 2, 3, 4, 6, 7, 12, 14, 21, 28, 42, 84
24) 1, 2, 5, 10, 11, 22, 55, 110
25) 1, 2, 17, 34
26) 1, 2, 7, 14, 49, 98
27) 1, 3, 7, 9, 21, 63
28) 1, 2, 53, 106

THEA Math Practice Book

Great Common Factor (GCF)

1) 3
2) 1
3) 2
4) 5
5) 4
6) 2
7) 5
8) 5
9) 9
10) 9
11) 2
12) 35
13) 10
14) 2
15) 48
16) 5
17) 8
18) 20
19) 1
20) 8
21) 5
22) 2
23) 15
24)

Least Common Multiple (LCM)

1) 40
2) 32
3) 30
4) 42
5) 60
6) 5
7) 120
8) 30
9) 702
10) 140
11) 54
12) 110
13) 60
14) 126
15) 40
16) 162
17) 38
18) 44
19) 300
20) 48
21) 170
22) 56
23) 70
24) 42

Divisibility Rule

1) 15   2, $\underline{3}$, 4, $\underline{5}$, 6, 9, 10
2) 124 $\underline{2}$, 3, $\underline{4}$, 5, 6, 9, 10
3) 352 $\underline{2}$, 3, $\underline{4}$, 5, 6, 9, 10
4) 94 $\underline{2}$, 3, 4, 5, 6, 9, 10
5) 241   2, 3, 4, 5, 6, 9, 10
6) 455   2, 3, 4, $\underline{5}$, 6, 9, 10
7) 65   2, 3, 4, $\underline{5}$, 6, 9, 10
8) 320 $\underline{2}$, 3, $\underline{4}$, $\underline{5}$, 6, 9, 10
9) 1,134 $\underline{2}$, $\underline{3}$, 4, $\underline{6}$, $\underline{9}$, 10
10) 68 $\underline{2}$, 3, $\underline{4}$, 5, 6, 9, 10
11) 754 $\underline{2}$, 3, 4, 5, 6, 9, 10
12) 148 $\underline{2}$, 3, $\underline{4}$, 5, 6, 9, 10
13) 34 $\underline{2}$, 3, 4, 5, 6, 9, 10
14) 385   2, 3, 4, $\underline{5}$, 6, 9, 10
15) 915   2, $\underline{3}$, 4, $\underline{5}$, 6, 9, 10
16) 157   2, 3, 4, 5, 6, 9, 10
17) 540 $\underline{2}$, $\underline{3}$, $\underline{4}$, $\underline{5}$, $\underline{6}$, $\underline{9}$, $\underline{10}$
18) 340 $\underline{2}$, 3, $\underline{4}$, $\underline{5}$, 6, 9, $\underline{10}$
19) 480 $\underline{2}$, $\underline{3}$, $\underline{4}$, $\underline{5}$, $\underline{6}$, 9, $\underline{10}$
20) 3,750 $\underline{2}$, $\underline{3}$, 4, $\underline{5}$, $\underline{6}$, 9, $\underline{10}$
21) 660 $\underline{2}$, $\underline{3}$, $\underline{4}$, $\underline{5}$, $\underline{6}$, 9, $\underline{10}$
22) 286 $\underline{2}$, 3, 4, 5, 6, 9, 10
23) 158 $\underline{2}$, 3, 4, 5, 6, 9, 10
24) 456 $\underline{2}$, $\underline{3}$, $\underline{4}$, 5, $\underline{6}$, 9, 10

Chapter 2:
Fundamental

Adding Fractions – Unlike Denominator

Add the fractions and simplify the answers.

1) $\frac{1}{4} + \frac{2}{3} =$

2) $\frac{4}{5} + \frac{1}{2} =$

3) $\frac{1}{4} + \frac{5}{7} =$

4) $\frac{8}{11} + \frac{1}{2} =$

5) $\frac{7}{18} + \frac{1}{3} =$

6) $\frac{13}{54} + \frac{5}{18} =$

7) $\frac{5}{8} + \frac{1}{6} =$

8) $\frac{3}{10} + \frac{1}{4} =$

9) $\frac{5}{11} + \frac{2}{4} =$

10) $\frac{1}{9} + \frac{4}{7} =$

11) $\frac{5}{18} + \frac{3}{8} =$

12) $\frac{7}{32} + \frac{3}{4} =$

13) $\frac{9}{65} + \frac{2}{5} =$

14) $\frac{8}{63} + \frac{3}{7} =$

15) $\frac{11}{64} + \frac{1}{4} =$

16) $\frac{4}{15} + \frac{2}{5} =$

17) $\frac{4}{7} + \frac{3}{6} =$

18) $\frac{5}{72} + \frac{2}{9} =$

19) $\frac{2}{15} + \frac{1}{25} =$

20) $\frac{5}{12} + \frac{3}{8} =$

21) $\frac{7}{88} + \frac{1}{8} =$

22) $\frac{7}{12} + \frac{2}{5} =$

23) $\frac{3}{72} + \frac{1}{4} =$

24) $\frac{2}{27} + \frac{1}{18} =$

Subtracting Fractions – Unlike Denominator

Solve each problem.

1) $\dfrac{3}{4} - \dfrac{1}{5} =$

2) $\dfrac{2}{3} - \dfrac{1}{4} =$

3) $\dfrac{5}{6} - \dfrac{3}{7} =$

4) $\dfrac{5}{6} - \dfrac{7}{12} =$

5) $\dfrac{6}{7} - \dfrac{3}{14} =$

6) $\dfrac{7}{12} - \dfrac{7}{18} =$

7) $\dfrac{17}{20} - \dfrac{2}{5} =$

8) $\dfrac{2}{3} - \dfrac{1}{16} =$

9) $\dfrac{6}{7} - \dfrac{4}{9} =$

10) $\dfrac{3}{8} - \dfrac{5}{32} =$

11) $\dfrac{5}{7} - \dfrac{4}{35} =$

12) $\dfrac{5}{6} - \dfrac{7}{30} =$

13) $\dfrac{6}{7} - \dfrac{4}{21} =$

14) $\dfrac{5}{3} - \dfrac{8}{15} =$

15) $\dfrac{2}{11} - \dfrac{3}{22} =$

16) $\dfrac{5}{6} - \dfrac{4}{54} =$

17) $\dfrac{7}{24} - \dfrac{7}{32} =$

18) $\dfrac{3}{4} - \dfrac{3}{5} =$

19) $\dfrac{1}{2} - \dfrac{2}{9} =$

20) $\dfrac{2}{3} - \dfrac{6}{11} =$

Converting Mix Numbers

Convert the following mixed numbers into improper fractions.

1) $3\frac{5}{6} =$

2) $5\frac{11}{15} =$

3) $4\frac{1}{3} =$

4) $2\frac{4}{7} =$

5) $7\frac{1}{4} =$

6) $3\frac{19}{21} =$

7) $5\frac{9}{10} =$

8) $4\frac{7}{12} =$

9) $3\frac{10}{11} =$

10) $6\frac{2}{5} =$

11) $8\frac{2}{3} =$

12) $2\frac{11}{12} =$

13) $3\frac{5}{6} =$

14) $4\frac{8}{11} =$

15) $7\frac{1}{4} =$

16) $5\frac{6}{11} =$

17) $8\frac{1}{5} =$

18) $3\frac{7}{12} =$

19) $6\frac{1}{22} =$

20) $3\frac{2}{3} =$

21) $7\frac{4}{5} =$

22) $4\frac{7}{8} =$

23) $6\frac{5}{6} =$

24) $12\frac{9}{10} =$

Converting improper Fractions

Convert the following improper fractions into mixed numbers

1) $\frac{62}{14}=$

2) $\frac{98}{37}=$

3) $\frac{49}{17}=$

4) $\frac{57}{23}=$

5) $\frac{71}{16}=$

6) $\frac{137}{42}=$

7) $\frac{120}{33}=$

8) $\frac{26}{5}=$

9) $\frac{33}{19}=$

10) $\frac{13}{2}=$

11) $\frac{39}{4}=$

12) $\frac{210}{65}=$

13) $\frac{76}{64}=$

14) $\frac{18}{7}=$

15) $\frac{110}{13}=$

16) $\frac{49}{4}=$

17) $\frac{122}{9}=$

18) $\frac{61}{12}=$

19) $\frac{37}{6}=$

20) $\frac{28}{9}=$

21) $\frac{5}{4}=$

22) $\frac{79}{13}=$

23) $\frac{41}{8}=$

24) $\frac{64}{7}=$

THEA Math Practice Book

Addition Mix Numbers

Add the following fractions.

1) $1\frac{1}{5} + 4\frac{2}{5} =$

2) $5\frac{3}{7} + 3\frac{4}{7} =$

3) $2\frac{2}{8} + 3\frac{1}{8} =$

4) $5\frac{5}{8} + 3\frac{1}{2} =$

5) $2\frac{9}{14} + 3\frac{3}{12} =$

6) $6\frac{2}{5} + 3\frac{1}{2} =$

7) $2\frac{8}{27} + 2\frac{2}{18} =$

8) $2\frac{3}{4} + 3\frac{1}{3} =$

9) $4\frac{5}{6} + 1\frac{1}{6} =$

10) $3\frac{5}{7} + 1\frac{3}{7} =$

11) $4\frac{1}{2} + 2\frac{2}{5} =$

12) $5\frac{1}{4} + 2\frac{5}{6} =$

13) $5\frac{1}{3} + 2\frac{2}{3} =$

14) $3\frac{5}{6} + 3\frac{2}{12} =$

15) $4\frac{3}{5} + 4\frac{1}{2} =$

16) $5\frac{2}{3} + 1\frac{4}{7} =$

17) $4\frac{5}{6} + 6\frac{1}{4} =$

18) $2\frac{2}{5} + 3\frac{3}{8} =$

19) $3\frac{1}{6} + 2\frac{4}{9} =$

20) $5\frac{3}{5} + 3\frac{2}{3} =$

21) $4\frac{5}{8} + 1\frac{1}{3} =$

22) $6\frac{1}{9} + 4\frac{4}{5} =$

23) $2\frac{2}{7} + 3\frac{4}{5} =$

24) $3\frac{1}{2} + 1\frac{5}{7} =$

Subtracting Mix Numbers

Subtract the following fractions.

1) $7\frac{1}{3} - 6\frac{1}{3} =$

2) $4\frac{5}{8} - 4\frac{2}{8} =$

3) $8\frac{5}{9} - 7\frac{1}{9} =$

4) $4\frac{1}{4} - 1\frac{1}{3} =$

5) $3\frac{1}{3} - 2\frac{1}{6} =$

6) $8\frac{1}{2} - 3\frac{2}{5} =$

7) $7\frac{5}{8} - 3\frac{3}{8} =$

8) $9\frac{9}{13} - 4\frac{6}{13} =$

9) $5\frac{7}{12} - 2\frac{5}{12} =$

10) $4\frac{4}{7} - 1\frac{3}{7} =$

11) $7\frac{1}{5} - 2\frac{1}{10} =$

12) $4\frac{5}{6} - 2\frac{1}{6} =$

13) $6\frac{2}{45} - 1\frac{1}{5} =$

14) $4\frac{1}{2} - 2\frac{1}{4} =$

15) $14\frac{4}{5} - 11\frac{2}{5} =$

16) $6\frac{2}{4} - 1\frac{1}{4} =$

17) $4\frac{1}{5} - 2\frac{3}{5} =$

18) $5\frac{1}{8} - 2\frac{1}{2} =$

19) $6\frac{2}{3} - 1\frac{1}{9} =$

20) $4\frac{3}{5} - 4\frac{1}{15} =$

21) $9\frac{9}{11} - 5\frac{1}{2} =$

22) $8\frac{4}{5} - 2\frac{3}{20} =$

23) $3\frac{2}{3} - 2\frac{1}{9} =$

24) $7\frac{9}{14} - 3\frac{3}{14} =$

Simplify Fractions

Reduce these fractions to lowest terms

1) $\frac{15}{10} =$

2) $\frac{20}{30} =$

3) $\frac{28}{35} =$

4) $\frac{21}{28} =$

5) $\frac{6}{18} =$

6) $\frac{27}{63} =$

7) $\frac{16}{28} =$

8) $\frac{48}{60} =$

9) $\frac{8}{72} =$

10) $\frac{30}{12} =$

11) $\frac{45}{60} =$

12) $\frac{30}{90} =$

13) $\frac{18}{30} =$

14) $\frac{5}{20} =$

15) $\frac{16}{56} =$

16) $\frac{56}{84} =$

17) $\frac{88}{33} =$

18) $\frac{36}{135} =$

19) $\frac{21}{56} =$

20) $\frac{64}{56} =$

21) $\frac{140}{280} =$

22) $\frac{30}{155} =$

23) $\frac{210}{42} =$

24) $\frac{130}{520} =$

Multiplying Fractions

Find the product.

1) $\dfrac{4}{5} \times \dfrac{2}{6} =$

2) $\dfrac{4}{22} \times \dfrac{5}{8} =$

3) $\dfrac{8}{30} \times \dfrac{12}{16} =$

4) $\dfrac{9}{14} \times \dfrac{21}{36} =$

5) $\dfrac{14}{15} \times \dfrac{5}{7} =$

6) $\dfrac{16}{19} \times \dfrac{3}{4} =$

7) $\dfrac{4}{9} \times \dfrac{9}{8} =$

8) $\dfrac{47}{85} \times 0 =$

9) $\dfrac{5}{8} \times \dfrac{16}{6} =$

10) $\dfrac{28}{15} \times \dfrac{5}{7} =$

11) $\dfrac{32}{24} \times \dfrac{12}{16} =$

12) $\dfrac{6}{42} \times \dfrac{7}{36} =$

13) $\dfrac{13}{8} \times \dfrac{12}{4} =$

14) $\dfrac{10}{9} \times \dfrac{6}{5} =$

15) $\dfrac{35}{56} \times \dfrac{8}{7} =$

16) $\dfrac{16}{18} \times 9 =$

17) $\dfrac{5}{22} \times \dfrac{44}{15} =$

18) $\dfrac{10}{18} \times \dfrac{9}{20} =$

19) $\dfrac{7}{11} \times \dfrac{8}{21} =$

20) $\dfrac{26}{24} \times \dfrac{8}{52} =$

21) $\dfrac{6}{17} \times \dfrac{1}{12} =$

22) $\dfrac{20}{9} \times \dfrac{6}{100} =$

23) $\dfrac{8}{14} \times \dfrac{7}{72} =$

24) $\dfrac{50}{100} \times \dfrac{300}{400} =$

Multiplying Mixed Number

Multiply. Reduce to lowest terms.

1) $2\frac{3}{5} \times 1\frac{3}{4} =$

2) $1\frac{5}{6} \times 1\frac{1}{3} =$

3) $2\frac{3}{5} \times 1\frac{1}{7} =$

4) $3\frac{1}{7} \times 2\frac{1}{2} =$

5) $4\frac{3}{4} \times 1\frac{1}{4} =$

6) $3\frac{1}{2} \times 1\frac{4}{5} =$

7) $3\frac{3}{4} \times 1\frac{1}{2} =$

8) $5\frac{2}{3} \times 3\frac{1}{3} =$

9) $3\frac{2}{3} \times 3\frac{1}{2} =$

10) $2\frac{1}{3} \times 3\frac{1}{2} =$

11) $4\frac{3}{4} \times 3\frac{2}{3} =$

12) $3\frac{2}{4} \times 3\frac{1}{6} =$

13) $2\frac{2}{5} \times 1\frac{1}{3} =$

14) $2\frac{1}{3} \times 1\frac{1}{6} =$

15) $2\frac{2}{3} \times 3\frac{1}{2} =$

16) $2\frac{1}{8} \times 2\frac{2}{5} =$

17) $2\frac{1}{4} \times 1\frac{2}{3} =$

18) $2\frac{3}{5} \times 1\frac{1}{4} =$

19) $2\frac{3}{5} \times 1\frac{5}{6} =$

20) $3\frac{3}{5} \times 2\frac{3}{4} =$

21) $3\frac{3}{4} \times 1\frac{1}{3} =$

22) $2\frac{5}{8} \times 3\frac{1}{4} =$

Dividing Fractions

Divide these fractions.

1) $1 \div \frac{1}{5} =$

2) $\frac{7}{13} \div 7 =$

3) $\frac{5}{14} \div \frac{2}{5} =$

4) $\frac{15}{60} \div \frac{3}{4} =$

5) $\frac{2}{17} \div \frac{4}{17} =$

6) $\frac{4}{16} \div \frac{18}{24} =$

7) $0 \div \frac{1}{9} =$

8) $\frac{12}{16} \div \frac{8}{9} =$

9) $\frac{8}{12} \div \frac{4}{18} =$

10) $\frac{9}{14} \div \frac{3}{7} =$

11) $\frac{8}{15} \div \frac{25}{16} =$

12) $\frac{35}{12} \div \frac{15}{6} =$

13) $\frac{9}{15} \div \frac{9}{5} =$

14) $\frac{8}{18} \div \frac{40}{6} =$

15) $\frac{45}{21} \div \frac{9}{21} =$

16) $\frac{7}{30} \div \frac{63}{5} =$

17) $\frac{36}{8} \div \frac{18}{24} =$

18) $9 \div \frac{1}{2} =$

19) $\frac{48}{35} \div \frac{8}{7} =$

20) $\frac{3}{36} \div \frac{9}{6} =$

21) $\frac{4}{7} \div \frac{12}{14} =$

22) $\frac{8}{40} \div \frac{10}{5} =$

Dividing Mixed Number

Divide the following mixed numbers. Cancel and simplify when possible.

1) $4\frac{1}{4} \div 4\frac{1}{3} =$

2) $2\frac{1}{6} \div 1\frac{2}{2} =$

3) $5\frac{1}{3} \div 3\frac{3}{4} =$

4) $3\frac{1}{6} \div 3\frac{1}{5} =$

5) $5\frac{1}{6} \div 1\frac{2}{3} =$

6) $3\frac{3}{5} \div 2\frac{2}{6} =$

7) $4\frac{3}{5} \div 2\frac{1}{3} =$

8) $2\frac{4}{9} \div 1\frac{1}{9} =$

9) $3\frac{5}{6} \div 3\frac{1}{2} =$

10) $9\frac{1}{9} \div 3\frac{2}{3} =$

11) $2\frac{2}{7} \div 4\frac{1}{7} =$

12) $4\frac{3}{8} \div 1\frac{3}{4} =$

13) $5\frac{1}{8} \div 1\frac{1}{12} =$

14) $6\frac{3}{8} \div 3\frac{1}{3} =$

15) $4\frac{2}{5} \div 1\frac{1}{5} =$

16) $2\frac{1}{2} \div 2\frac{2}{9} =$

17) $7\frac{1}{6} \div 5\frac{3}{8} =$

18) $5\frac{1}{2} \div 4\frac{1}{3} =$

19) $4\frac{5}{7} \div 1\frac{1}{3} =$

20) $3\frac{5}{6} \div 1\frac{1}{4} =$

21) $9\frac{1}{2} \div 7\frac{1}{3} =$

22) $3\frac{1}{8} \div 1\frac{1}{9} =$

23) $4\frac{1}{4} \div 3\frac{3}{4} =$

24) $3\frac{1}{6} \div 3\frac{1}{3} =$

Comparing Fractions

Compare the fractions, and write >, < or =

1) $\dfrac{14}{3}$ ____ $\dfrac{24}{15}$

2) $\dfrac{32}{3}$ ____ $\dfrac{2}{5}$

3) $\dfrac{4}{9}$ ____ $\dfrac{2}{4}$

4) $\dfrac{12}{4}$ ____ $\dfrac{13}{9}$

5) $\dfrac{1}{8}$ ____ $\dfrac{2}{3}$

6) $\dfrac{10}{6}$ ____ $\dfrac{16}{7}$

7) $\dfrac{12}{13}$ ____ $\dfrac{7}{9}$

8) $\dfrac{20}{14}$ ____ $\dfrac{25}{3}$

9) $4\dfrac{1}{12}$ ____ $6\dfrac{1}{3}$

10) $8\dfrac{1}{6}$ ____ $3\dfrac{1}{8}$

11) $3\dfrac{1}{2}$ ____ $3\dfrac{1}{5}$

12) $7\dfrac{5}{8}$ ____ $7\dfrac{2}{9}$

13) $3\dfrac{2}{8}$ ____ $5\dfrac{3}{5}$

14) $\dfrac{1}{15}$ ____ $\dfrac{3}{7}$

15) $\dfrac{31}{25}$ ____ $\dfrac{19}{83}$

16) $\dfrac{12}{100}$ ____ $\dfrac{6}{62}$

17) $15\dfrac{1}{4}$ ____ $15\dfrac{1}{9}$

18) $\dfrac{1}{5}$ ____ $\dfrac{1}{9}$

19) $\dfrac{1}{7}$ ____ $\dfrac{1}{13}$

20) $\dfrac{1}{18}$ ____ $\dfrac{8}{15}$

21) $\dfrac{7}{22}$ ____ $\dfrac{9}{76}$

22) $\dfrac{4}{5}$ ____ $\dfrac{2}{5}$

23) $2\dfrac{17}{14}$ ____ $3\dfrac{3}{14}$

24) $3\dfrac{25}{4}$ ____ $4\dfrac{5}{4}$

Answer key Chapter 2

Adding Fractions – Unlike Denominator

1) $\frac{11}{12}$
2) $\frac{13}{10}$
3) $\frac{27}{28}$
4) $\frac{27}{22}$
5) $\frac{13}{18}$
6) $\frac{14}{27}$
7) $\frac{19}{24}$
8) $\frac{11}{20}$
9) $\frac{21}{22}$
10) $\frac{43}{63}$
11) $\frac{47}{72}$
12) $\frac{31}{32}$
13) $\frac{7}{13}$
14) $\frac{5}{9}$
15) $\frac{27}{64}$
16) $\frac{2}{3}$
17) $\frac{15}{14}$
18) $\frac{7}{24}$
19) $\frac{13}{75}$
20) $\frac{19}{24}$
21) $\frac{9}{44}$
22) $\frac{59}{60}$
23) $\frac{7}{24}$
24) $\frac{7}{54}$

Subtracting Fractions – Unlike Denominator

1) $\frac{11}{20}$
2) $\frac{5}{12}$
3) $\frac{17}{42}$
4) $\frac{1}{4}$
5) $\frac{9}{14}$
6) $\frac{7}{36}$
7) $\frac{9}{20}$
8) $\frac{29}{48}$
9) $\frac{26}{63}$
10) $\frac{7}{32}$
11) $\frac{3}{5}$
12) $\frac{3}{5}$
13) $\frac{2}{3}$
14) $\frac{17}{15}$
15) $\frac{1}{22}$
16) $\frac{41}{54}$
17) $\frac{7}{96}$
18) $\frac{3}{20}$
19) $\frac{5}{18}$
20) $\frac{4}{33}$

Converting Mix Numbers

1) $\frac{23}{6}$
2) $\frac{86}{15}$
3) $\frac{13}{3}$
4) $\frac{18}{7}$
5) $\frac{29}{4}$
6) $\frac{82}{21}$
7) $\frac{59}{10}$
8) $\frac{55}{12}$
9) $\frac{43}{11}$
10) $\frac{32}{5}$
11) $\frac{26}{3}$
12) $\frac{35}{12}$
13) $\frac{23}{6}$
14) $\frac{52}{11}$
15) $\frac{29}{4}$
16) $\frac{61}{11}$
17) $\frac{41}{5}$
18) $\frac{43}{12}$

19) $\frac{133}{22}$

20) $\frac{11}{3}$

21) $\frac{39}{5}$

22) $\frac{39}{8}$

23) $\frac{41}{6}$

24) $\frac{129}{10}$

Converting improper Fractions

1) $4\frac{6}{14}$
2) $2\frac{24}{37}$
3) $2\frac{15}{17}$
4) $2\frac{11}{23}$
5) $4\frac{7}{16}$
6) $3\frac{11}{42}$
7) $3\frac{21}{33}$
8) $5\frac{1}{5}$

9) $1\frac{14}{19}$
10) $6\frac{1}{2}$
11) $9\frac{3}{4}$
12) $3\frac{15}{65}$
13) $1\frac{12}{64}$
14) $2\frac{4}{7}$
15) $8\frac{6}{13}$
16) $12\frac{1}{4}$

17) $13\frac{5}{9}$
18) $5\frac{1}{12}$
19) $6\frac{1}{6}$
20) $3\frac{1}{9}$
21) $1\frac{1}{4}$
22) $6\frac{1}{13}$
23) $5\frac{1}{8}$
24) $9\frac{1}{7}$

Adding Mix Numbers

1) $5\frac{3}{5}$
2) 9
3) $5\frac{3}{8}$
4) $9\frac{1}{8}$
5) $5\frac{25}{28}$
6) $9\frac{9}{10}$
7) $4\frac{11}{27}$
8) $6\frac{1}{12}$

9) 6
10) $5\frac{1}{7}$
11) $6\frac{9}{10}$
12) $8\frac{1}{12}$
13) 8
14) 7
15) $9\frac{1}{10}$
16) $7\frac{5}{21}$

17) $11\frac{1}{12}$
18) $5\frac{31}{40}$
19) $5\frac{11}{18}$
20) $9\frac{4}{15}$
21) $5\frac{23}{24}$
22) $10\frac{41}{45}$
23) $6\frac{3}{35}$
24) $5\frac{3}{14}$

Subtracting Mix Numbers

1) 1
2) $\frac{3}{8}$
3) $1\frac{4}{9}$
4) $2\frac{11}{12}$

5) $1\frac{1}{6}$
6) $5\frac{1}{10}$
7) $4\frac{1}{4}$

8) $5\frac{3}{13}$
9) $3\frac{1}{6}$
10) $3\frac{1}{7}$

11) $5\frac{1}{10}$
12) $2\frac{2}{3}$
13) $4\frac{38}{45}$
14) $2\frac{1}{4}$
15) $3\frac{2}{5}$

16) $5\frac{1}{4}$
17) $1\frac{3}{5}$
18) $2\frac{5}{8}$
19) $5\frac{5}{9}$
20) $\frac{8}{15}$

21) $4\frac{7}{22}$
22) $6\frac{13}{20}$
23) $1\frac{5}{9}$
24) $4\frac{3}{7}$

Simplify Fractions

1) $\frac{3}{2}$
2) $\frac{2}{3}$
3) $\frac{4}{5}$
4) $\frac{3}{4}$
5) $\frac{1}{3}$
6) $\frac{3}{7}$
7) $\frac{4}{7}$
8) $\frac{4}{5}$

9) $\frac{1}{9}$
10) $\frac{5}{2}$
11) $\frac{3}{4}$
12) $\frac{1}{3}$
13) $\frac{3}{5}$
14) $\frac{1}{4}$
15) $\frac{2}{7}$
16) $\frac{2}{3}$

17) $\frac{8}{3}$
18) $\frac{4}{15}$
19) $\frac{3}{8}$
20) $\frac{8}{7}$
21) $\frac{1}{2}$
22) $\frac{6}{31}$
23) 5
24) $\frac{1}{4}$

Multiplying Fractions

1) $\frac{4}{15}$
2) $\frac{5}{44}$
3) $\frac{1}{5}$
4) $\frac{3}{8}$
5) $\frac{2}{3}$
6) $\frac{12}{19}$
7) $\frac{1}{2}$
8) 0

9) $\frac{5}{3}$
10) $\frac{4}{3}$
11) 1
12) $\frac{1}{36}$
13) $\frac{39}{8}$
14) $\frac{4}{3}$
15) $\frac{5}{7}$
16) 8

17) $\frac{2}{3}$
18) $\frac{1}{4}$
19) $\frac{8}{33}$
20) $\frac{1}{6}$
21) $\frac{1}{34}$
22) $\frac{2}{15}$
23) $\frac{1}{18}$
24) $\frac{3}{8}$

Multiplying Mixed Number

1) $4\frac{11}{20}$
2) $2\frac{4}{9}$
3) $2\frac{34}{35}$

THEA Math Practice Book

4) $7\frac{6}{7}$

5) $5\frac{15}{16}$

6) $6\frac{3}{10}$

7) $5\frac{5}{8}$

8) $18\frac{8}{9}$

9) $12\frac{5}{6}$

10) $8\frac{1}{6}$

11) $17\frac{5}{12}$

12) $11\frac{1}{12}$

13) $3\frac{1}{5}$

14) $2\frac{13}{18}$

15) $9\frac{1}{3}$

16) $5\frac{1}{10}$

17) $3\frac{3}{4}$

18) $3\frac{1}{4}$

19) $4\frac{23}{30}$

20) $9\frac{9}{10}$

21) 5

22) $8\frac{17}{32}$

Dividing Fractions

1) 5

2) $\frac{1}{13}$

3) $\frac{25}{28}$

4) $\frac{1}{3}$

5) $\frac{1}{2}$

6) $\frac{1}{3}$

7) 0

8) $\frac{27}{32}$

9) 3

10) $\frac{3}{2}$

11) $\frac{128}{375}$

12) $\frac{7}{6}$

13) $\frac{1}{3}$

14) $\frac{1}{15}$

15) 5

16) $\frac{1}{54}$

17) 6

18) 18

19) $\frac{6}{5}$

20) $\frac{1}{18}$

21) $\frac{2}{3}$

22) $\frac{1}{10}$

Dividing Mixed Number

1) $\frac{51}{52}$

2) $\frac{1}{12}$

3) $1\frac{19}{45}$

4) $\frac{95}{96}$

5) $3\frac{1}{10}$

6) $1\frac{19}{35}$

7) $1\frac{34}{35}$

8) $\frac{2}{81}$

9) $1\frac{2}{21}$

10) $2\frac{16}{33}$

11) $\frac{16}{29}$

12) $2\frac{1}{2}$

13) $4\frac{19}{26}$

14) $1\frac{73}{80}$

15) $3\frac{2}{3}$

16) $1\frac{1}{8}$

17) $1\frac{1}{3}$

18) $1\frac{7}{26}$

19) $3\frac{15}{28}$

20) $3\frac{1}{15}$

21) $1\frac{13}{44}$

22) $2\frac{13}{16}$ 23) $1\frac{2}{15}$ 24) $\frac{19}{20}$

Comparing Fractions

1) >	7) >	13) <	19) >
2) >	8) <	14) <	20) <
3) <	9) <	15) >	21) >
4) >	10) >	16) <	22) >
5) <	11) >	17) >	23) =
6) <	12) >	18) >	24) >

Chapter 3:

Decimal

Round Decimals

Round each number to the correct place value

1) 0.8<u>3</u> =

2) 3.<u>0</u>2 =

3) 7.<u>7</u>11 =

4) 0.<u>4</u>78 =

5) <u>8</u>.824 =

6) 0.0<u>7</u>8 =

7) 8.<u>1</u>3 =

8) 84.8<u>4</u>0 =

9) 2.5<u>3</u>8 =

10) 12.<u>2</u>97 =

11) 2.<u>0</u>8 =

12) 5.<u>3</u>24 =

13) 2.<u>1</u>32 =

14) 8.0<u>7</u>32 =

15) 5<u>5</u>.78 =

16) 2<u>8</u>.24 =

17) 5<u>2</u>7.156 =

18) 624.<u>7</u>88 =

19) 17.4<u>8</u>1 =

20) 9<u>4</u>.86 =

21) 4.3<u>0</u>67 =

22) 57.<u>0</u>86 =

23) 224.<u>2</u>24 =

24) 0.1<u>3</u>44 =

25) 0.00<u>6</u>9 =

26) 9.0<u>3</u>86 =

27) 35.5<u>4</u>22 =

28) 11.0<u>9</u>31 =

Decimals Addition

Add the following.

1) 32.12 + 24.28

2) 0.88 + 0.21

3) 15.36 + 10.87

4) 75.165 + 4.105

5) 8.650 + 7.82

6) 5.324 + 2.138

7) 81.21 + 15.85

8) 56.25 + 22.35

9) 46.21 + 10.07

10) 8.96 + 11.23

11) 15.214 + 11.251

12) 72.36 + 5.32

13) 32.05 + 8.54

14) 137.21 + 2.75

Decimals Subtraction

Subtract the following

1) 9.35 − 3.52

2) 75.35 − 62.37

3) 0.68 − 0.4

4) 11.245 − 8.6

5) 0.652 − 0.09

6) 75.25 − 28.88

7) 112.66 − 88.98

8) 32.56 − 12.45

9) 68.35 − 59.98

10) 6.985 − 0.223

11) 55.69 − 45.32

12) 12.352 − 2.325

13) 19.231 − 4.128

14) 128.98 − 7.92

Decimals Multiplication

Solve.

1) 3.1 × 3.4

2) 7.5 × 4.5

3) 5.04 × 3.04

4) 88.09 × 100

5) 23.9 × 10

6) 35.62 × 5.5

7) 32.75 × 11.3

8) 2.65 × 8.35

9) 12.05 × 0.04

10) 24.04 × 8.08

11) 12.34 × 11.2

12) 6.37 × 0.02

13) 9.4 × 0.14

14) 15.4 × 6.05

Decimal Division

Dividing Decimals.

1) 8 ÷ 10,000 =

2) 4 ÷ 100 =

3) 3.4 ÷ 100 =

4) 0.002 ÷ 10 =

5) 8 ÷ 64 =

6) 3 ÷ 81 =

7) 5 ÷ 45 =

8) 9 ÷ 180 =

9) 7 ÷ 1,000 =

10) 0.6 ÷ 0.63 =

11) 0.9 ÷ 0.009 =

12) 0.6 ÷ 0.12 =

13) 0.6 ÷ 0.42 =

14) 0.4 ÷ 0.04 =

15) 3.08 ÷ 10 =

16) 9.4 ÷ 10 =

17) 6.75 ÷ 100 =

18) 18.3 ÷ 3.3 =

19) 64.4 ÷ 4 =

20) 0.4 ÷ 0.004 =

21) 7.05 ÷ 3.5 =

22) 0.08 ÷ 0.40 =

23) 0.9 ÷ 7.6 =

24) 0.09 ÷ 54 =

25) 5.24 ÷ 0.5 =

26) 0.025 ÷ 125 =

Comparing Decimals

Write the Correct Comparison Symbol (>, < or =)

1) 1.42 ____ 2.42

2) 0.5 ____ 0.425

3) 13.6 ____ 13.600

4) 7.07 ____ 7.70

5) 0.922 ____ 0.92

6) 0.856 ____ 0.956

7) 4.34 ____ 4.242

8) 5.0025 ____ 5.025

9) 24.087 ____ 24.078

10) 7.12 ____ 7.29

11) 4.44 ____ 4.444

12) 0.09 ____ 0.18

13) 1.302 ____ 1.32

14) 9.56 ____ 9.0569

15) 0.33 ____ 0.033

16) 21.04 ____ 21.040

17) 0.250 ____ 0.35

18) 44.92 ____ 45.01

19) 0.085 ____ 0.805

20) 36.5 ____ 29.8

21) 7.89 ____ 10.2

22) 0.024 ____ 0.0204

23) 5.042 ____ 0.5042

24) 7.5 ____ 0.758

25) 6.5 ____ 0.659

26) 3.24 ____ 3.2400

27) 8.34 ____ 0.834

28) 2.0809 ____ 2.0890

Convert Fraction to Decimal

Write each as a decimal.

1) $\frac{50}{100} =$

2) $\frac{46}{100} =$

3) $\frac{8}{50} =$

4) $\frac{8}{32} =$

5) $\frac{8}{72} =$

6) $\frac{56}{100} =$

7) $\frac{4}{50} =$

8) $\frac{31}{48} =$

9) $\frac{27}{300} =$

10) $\frac{15}{55} =$

11) $\frac{16}{32} =$

12) $\frac{6}{16} =$

13) $\frac{3}{10} =$

14) $\frac{18}{250} =$

15) $\frac{24}{80} =$

16) $\frac{30}{40} =$

17) $\frac{68}{100} =$

18) $\frac{7}{35} =$

19) $\frac{87}{100} =$

20) $\frac{1}{100} =$

21) $\frac{6}{36} =$

22) $\frac{2}{80} =$

Convert Decimal to Percent

Write each as a percent.

1) 0.187 =

2) 0.19 =

3) 2.6 =

4) 0.017 =

5) 0.009 =

6) 0.786 =

7) 0.245 =

8) 0.57 =

9) 0.002 =

10) 0.205 =

11) 0.324 =

12) 84.9 =

13) 3.015 =

14) 0.7 =

15) 2.35 =

16) 0.0367 =

17) 0.0043 =

18) 0.960 =

19) 6.68 =

20) 0.484 =

21) 8.957 =

22) 0.879 =

23) 2.7 =

24) 0.9 =

25) 3.6 =

26) 26.8 =

27) 1.01 =

28) 0.006 =

Convert Fraction to Percent

Write each as a percent.

1) $\frac{1}{4} =$

2) $\frac{3}{8} =$

3) $\frac{7}{14} =$

4) $\frac{15}{35} =$

5) $\frac{12}{28} =$

6) $\frac{17}{68} =$

7) $\frac{8}{11} =$

8) $\frac{14}{30} =$

9) $\frac{6}{50} =$

10) $\frac{12}{48} =$

11) $\frac{5}{34} =$

12) $\frac{27}{10} =$

13) $\frac{24}{80} =$

14) $\frac{16}{25} =$

15) $\frac{16}{58} =$

16) $\frac{2}{22} =$

17) $\frac{32}{88} =$

18) $\frac{21}{36} =$

19) $\frac{18}{92} =$

20) $\frac{6}{60} =$

21) $\frac{24}{600} =$

22) $\frac{720}{360} =$

Answer key Chapter 3

Round Decimals
1) 0.8
2) 3.0
3) 7.7
4) 0.5
5) 9.0
6) 0.08
7) 8.1
8) 84.84
9) 2.54
10) 12.3
11) 2.1
12) 5.3
13) 2.1
14) 8.07
15) 56.0
16) 28.0
17) 530.0
18) 624.8
19) 17.48
20) 95.0
21) 4.31
22) 57.1
23) 224.2
24) 0.13
25) 0.007
26) 9.04
27) 35.54
28) 11.09

Decimals Addition
1) 56.4
2) 1.09
3) 26.23
4) 79.27
5) 16.47
6) 7.462
7) 97.06
8) 78.6
9) 56.28
10) 20.19
11) 26.465
12) 77.68
13) 40.59
14) 139.96

Decimals Subtraction
1) 5.83
2) 12.98
3) 0.28
4) 2.645
5) 0.562
6) 46.37
7) 23.68
8) 20.11
9) 8.37
10) 6.762
11) 10.37
12) 10.027
13) 15.103
14) 121.06

Decimals Multiplication
1) 10.54
2) 33.75
3) 15.3216
4) 8,809
5) 239
6) 195.91
7) 370.075
8) 22.1275
9) 0.482
10) 194.2432
11) 138.208
12) 0.1274
13) 1.316
14) 93.17

Decimal Division
1) 0.0008
2) 0.04
3) 0.034

THEA Math Practice Book

4) 0.0002
5) 0.125
6) 0.037....
7) 0.111...
8) 0.05
9) 0.007
10) 0.952...
11) 100

12) 5
13) 1.4285...
14) 10
15) 0.308
16) 0.94
17) 0.0675
18) 5.5454...
19) 16.1

20) 100
21) 2.01428...
22) 0.2
23) 0.1184...
24) 0.0016
25) 10.48
26) 0.0002

Comparing Decimals

1) <
2) >
3) =
4) <
5) >
6) <
7) >
8) <
9) >
10) >

11) <
12) <
13) <
14) >
15) >
16) =
17) <
18) <
19) <
20) >

21) <
22) >
23) >
24) >
25) >
26) =
27) >
28) <

Convert Fraction to Decimal

1) 0.5
2) 0.46
3) 0.16
4) 0.25
5) 0.11
6) 0.56
7) 0.08
8) 0.646

9) 0.09
10) 0.27
11) 0.5
12) 0.375
13) 0.3
14) 0.072
15) 0.3
16) 0.75

17) 0.68
18) 0.2
19) 0.87
20) 0.01
21) 0.166
22) 0.025

Convert Decimal to Percent

1) 18.7%
2) 19%
3) 260%

4) 1.7%
5) 0.9%
6) 78.6%

7) 24.5%
8) 57%
9) 0.2%

10) 20.5%
11) 32.4%
12) 8,490%
13) 301.5%
14) 70%
15) 235%
16) 3.67%

17) 0.43%
18) 96%
19) 668%
20) 48.4%
21) 895.7%
22) 87.9%
23) 270%

24) 90%
25) 360%
26) 2,680%
27) 101%
28) 0.6%

Convert Fraction to Percent

1) 25%
2) 37.5%
3) 50%
4) 42.86%
5) 29.31%
6) 25%
7) 72.72%
8) 46.66%

9) 12%
10) 25%
11) 14.7%
12) 2.7%
13) 30%
14) 64%
15) 27.58%
16) 9.09%

17) 36.36%
18) 58.33%
19) 19.56%
20) 10%
21) 4%
22) 200%

Chapter 4: Equations and Inequality

Distributive and Simplifying Expressions

Simplify each expression.

1) $6x + 2 - 8 =$

2) $-(-4 - 5x) =$

3) $(-3x + 4)(-2) =$

4) $(-2x)(x + 3) =$

5) $-2x + x^2 + 4x^2 =$

6) $7y + 7x + 8y - 5x =$

7) $-3x + 3y + 14x - 9y =$

8) $-2x - 5 + 8x + \frac{16}{4} =$

9) $5 - 8(x - 2) =$

10) $-5 - 5x + 3x =$

11) $(x - 3y)2 + 4y =$

12) $2.5x^2 \times (-5x) =$

13) $-4 - 2x^2 + 6x^2 =$

14) $8 + 14x^2 + 4 =$

15) $4(-2x - 7) + 10 =$

16) $(-x)(-2 + 3x) - x(7 + x) =$

17) $-3(6 + 12) - 3x + 5x =$

18) $-4(5 - 12x - 3x) =$

19) $3(-2x - 6) =$

20) $9 + 7x - 9 =$

21) $x(-2x + 8) =$

22) $5xy + 4x - 3y + x + 2y =$

23) $3(-x - 7) + 9 =$

24) $(-3x - 4) + 7 =$

25) $3x + 4y - 5 + 1 =$

26) $(-2 + 3x) - 3x(1 + 2x) =$

27) $(-3)(-3x - 3y) =$

28) $4(-x - 2) + 5 =$

Factoring Expressions

Factor the common factor out of each expression.

1) $12x - 6 =$

2) $5x - 15 =$

3) $\frac{45}{15}x - 15 =$

4) $7b - 28 =$

5) $4a^2 - 24a =$

6) $2xy - 10y =$

7) $5x^2y + 15x =$

8) $a^2 - 8a + 7ab =$

9) $2a^2 + 2ab =$

10) $4x + 20 =$

11) $24x - 36xy =$

12) $8x - 6 =$

13) $\frac{1}{4}x - \frac{3}{4}y =$

14) $7xy - \frac{14}{3}x =$

15) $3ab + 9c =$

16) $\frac{1}{3}x - \frac{4}{3} =$

17) $10x - 15xy =$

18) $x^2 + 8x =$

19) $4x^2 - 12y =$

20) $4x^3 + 3xy + x^2 =$

21) $21x - 14 =$

22) $20b - 60c + 20d =$

23) $24ab - 8ac =$

24) $ax - ay - 3x + 3y =$

25) $3ax + 4a + 9x + 12 =$

26) $x^2 - 10x =$

27) $9x^3 - 18x^2 =$

28) $5x^2 - 70xy =$

Evaluate One Variable Expressions

Evaluate each using the values given.

1) $x + 4x, x = 3$

2) $5(-6 + 3x), x = 1$

3) $4x + 7x, x = -3$

4) $5(2 - x) + 5, x = 3$

5) $6x + 4x - 10, x = 2$

6) $5x + 11x + 12, x = -1$

7) $5x - 2x - 4, x = 5$

8) $\frac{3(5x+8)}{9}, x = 2$

9) $2x - 85, x = 32$

10) $\frac{x}{18}, x = 108$

11) $7(3 + 2x) - 33, x = 5$

12) $7(x + 3) - 23, x = 4$

13) $\frac{x+(-6)}{-3}, x = -6$

14) $8(6 - 3x) + 5, x = 2$

15) $-11 - \frac{x}{5} + 3x, x = 10$

16) $5x + 11x, x = 1$

17) $-12x + 3(5 + 3x), x = -7$

18) $x + 11x, x = 0.5$

19) $\frac{(2x-2)}{6}, x = 13$

20) $3(-1 - 2x), x = 5$

21) $5x - (5 - x), x = 3$

22) $\left(-\frac{15}{x}\right) + 2 + x, x = 5$

23) $-\frac{x \times 5}{x}, x = 5$

24) $2(-1 - 3x), x = 2$

25) $2x^2 + 7x, x = 1$

26) $2(3x + 1) - 4(x - 5), x = 3$

27) $-6x - 4, x = -5$

28) $7x + 2x, x = 3$

THEA Math Practice Book

Evaluate Two Variable Expressions

Evaluate the expressions.

1) $x + 4y$, $x = 5, y = 2$

2) $(-2)(-3x - 2y)$, $x = 1, y = 2$

3) $4x + 2y$, $x = 10, y = 5$

4) $\dfrac{x-4}{y+1}$, $x = 8, y = 3$

5) $\dfrac{a}{4} - 6b$, $a = 32, b = 4$

6) $3x - 4(y - 8)$, $x = 5, y = 3$

7) $3x + 2y - 10$, $x = 2, y = 10$

8) $-3x + 10 + 8y - 5$, $x = 2, y = 1$

9) $yx \div 3$, $x = 9, y = 9$

10) $a - b \div 3$, $a = 3, b = 12$

11) $6(x - y)$, $x = 7, y = 4$

12) $5x - 4y$, $x = 5, y = 8$

13) $\dfrac{10}{a} + 3b$, $a = 5, b = 4$

14) $2x^2 + 4xy$, $x = 3, y = 5$

15) $8 - \dfrac{xy}{10} + y$, $x = 6, y = 5$

16) $7(3x - y)$, $x = 7, y = -9$

17) $5x^2 - 3y^2$, $x = -1, y = 2$

18) $3x + \dfrac{y}{4}$, $x = 6, y = 16$

19) $4(4x - 2y)$, $x = 3, y = 5$

20) $4x(y - \dfrac{1}{2})$, $x = 5, y = 4$

21) $5(x^2 - 2y)$, $x = 3, y = 2$

22) $5xy$, $x = 2, y = 8$

23) $\dfrac{1}{3}y^3\left(y - \dfrac{1}{4}x\right)$, $x = -4, y = 3$

24) $-3(x - 5y) - 2x$, $x = 4, y = 2$

25) $-2x + \dfrac{1}{6}xy$, $x = 3, y = 6$

26) $x^2 + xy^2$, $x = 5, y = 7$

27) $x - 2y + 8$, $x = 9, y = 6$

28) $\dfrac{xy}{2x+y}$, $x = 5, y = 4$

WWW.MathNotion.com

Graphing Linear Equation

Sketch the graph of each line.

1) $y = 2x - 5$
2) $y = -2x + 3$
3) $x - y = 0$

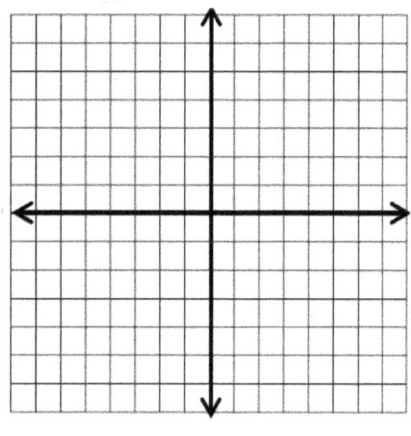
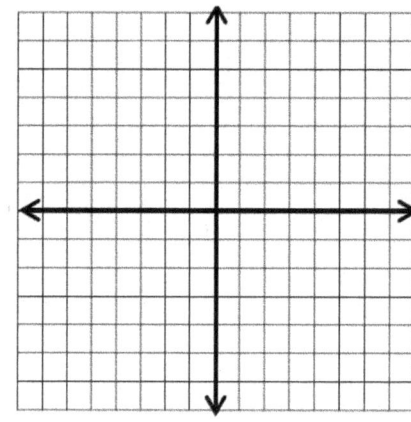
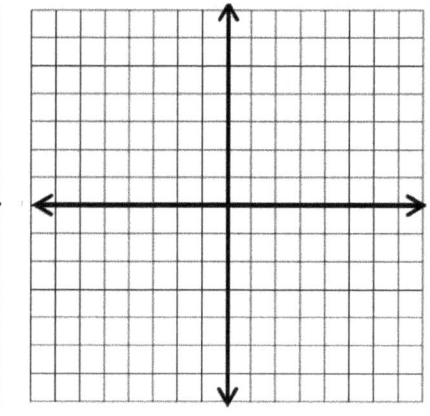

4) $x + y = 3$
5) $5x + 3y = -2$
6) $y - 3x + 2 = 0$

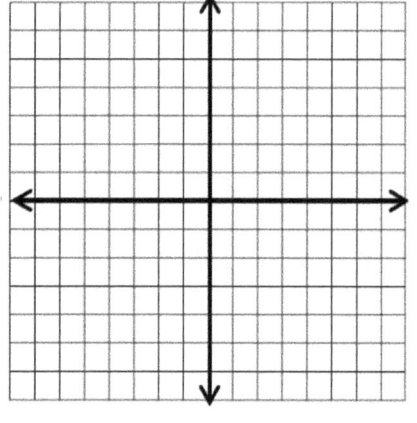
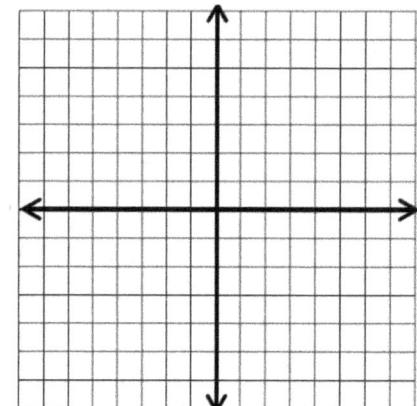
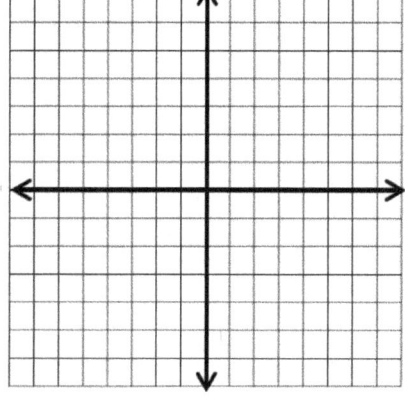

One Step Equations

Solve each equation.

1) $44 = (-12) + x$

2) $8x = (-64)$

3) $(-72) = (-8x)$

4) $(-5) = 3 + x$

5) $4 + \frac{x}{2} = (-3)$

6) $8x = (-104)$

7) $62 = x - 13$

8) $\frac{x}{3} = (-15)$

9) $x + 112 = 154$

10) $x - \frac{1}{3} = \frac{2}{3}$

11) $(-24) = x - 32$

12) $(-3x) = 39$

13) $(-169) = (13x)$

14) $-4x + 42 = 50$

15) $5x + 3 = 38$

16) $80 = (-8x)$

17) $3x + 7 = 19$

18) $24x = 144$

19) $x - 18 = 15$

20) $0.9x = 4.5$

21) $4x = 84$

22) $2x + 2.98 = 66.98$

23) $x + 9 = 6$

24) $x + 14 = 6$

25) $9x + 41 = 5$

26) $\frac{1}{4}x + 30 = 12$

Two Steps Equations

Solve each equation.

1) $6(3 + x) = 42$

2) $(-7)(x - 2) = 56$

3) $(-8)(3x - 4) = (-16)$

4) $5(2 + x) = -15$

5) $19(3x + 11) = 38$

6) $4(2x + 2) = 24$

7) $5(8 + 3x) = (-20)$

8) $(-5)(5x - 3) = 40$

9) $2x + 12 = 16$

10) $\frac{4x - 5}{5} = 3$

11) $(-3) = \frac{x + 4}{7}$

12) $80 = (-8)(x - 3)$

13) $\frac{x}{3} + 7 = 19$

14) $\frac{1}{4} = \frac{1}{2} + \frac{x}{4}$

15) $\frac{11 + x}{5} = (-6)$

16) $(-3)(10 + 5x) = (-15)$

17) $(-3x) + 12 = 24$

18) $\frac{x + 5}{5} = -5$

19) $\frac{x + 23}{8} = 3$

20) $(-4) + \frac{x}{2} = (-14)$

21) $-5 = \frac{x + 7}{8}$

22) $\frac{9x - 3}{6} = 4$

23) $\frac{2x - 12}{8} = 6$

24) $40 = (-5)(x - 8)$

Multi Steps Equations

Solve each equation.

1) $2 - (4 - 5x) = 3$

2) $-15 = -(4x + 7)$

3) $6x - 18 = (-2x) + 6$

4) $-32 = (-5x) - 11x$

5) $3(2 + 3x) + 3x = -30$

6) $5x - 18 = 2 + 2x - 7 + 2x$

7) $12 - 6x = (-36) - 3x + 3x$

8) $16 - 4x - 4x = 8 - 4x$

9) $8 + 7x + x = (-12) + 3x$

10) $(-3x) - 3(-2 + 4x) = 366$

11) $20 = (-200x) - 5 + 5$

12) $61 = 5x - 23 + 7x$

13) $7(4 + 2x) = 140$

14) $-60 = (-7x) - 13x$

15) $2(4x + 5) = -2(x + 4) - 22$

16) $11x - 17 = 6x + 8$

17) $9 = -3(x - 8)$

18) $(-6) - 8x = 6(1 + 2x)$

19) $x + 3 = -2(9 + 3x)$

20) $10 = 4 - 5x - 9$

21) $-15 - 9x - 3x = 12 - 3x$

22) $-23 - 3x + 5x = 27 - 23x$

23) $19 - 6x - 9x = -5 - 9x$

24) $15x - 18 = 6x + 9$

Graphing Linear Inequalities

Sketch the graph of each linear inequality.

1) $y > 2x - 3$ 2) $y < x + 3$ 3) $y \leq -3x - 8$

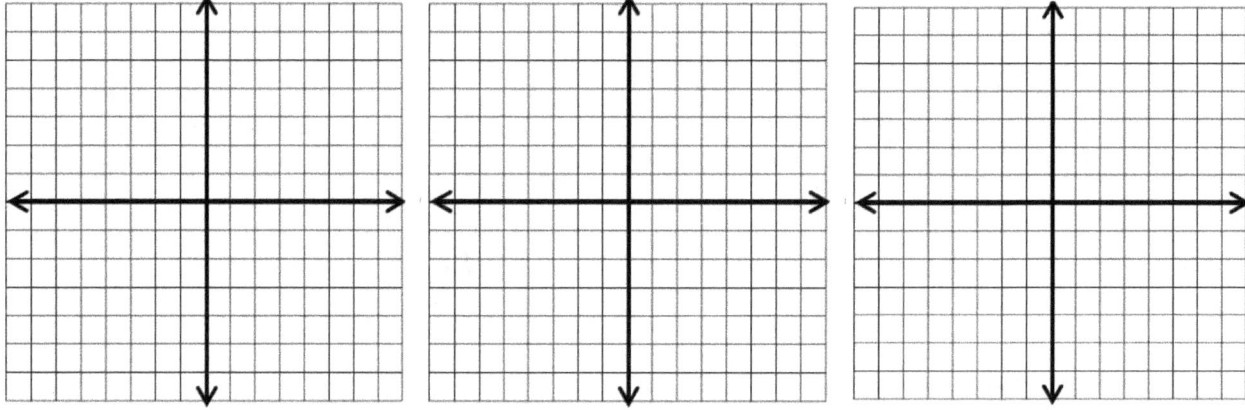

4) $3y \geq 6 + 3x$ 5) $-3y < x - 12$ 6) $2y \geq -8x + 4$

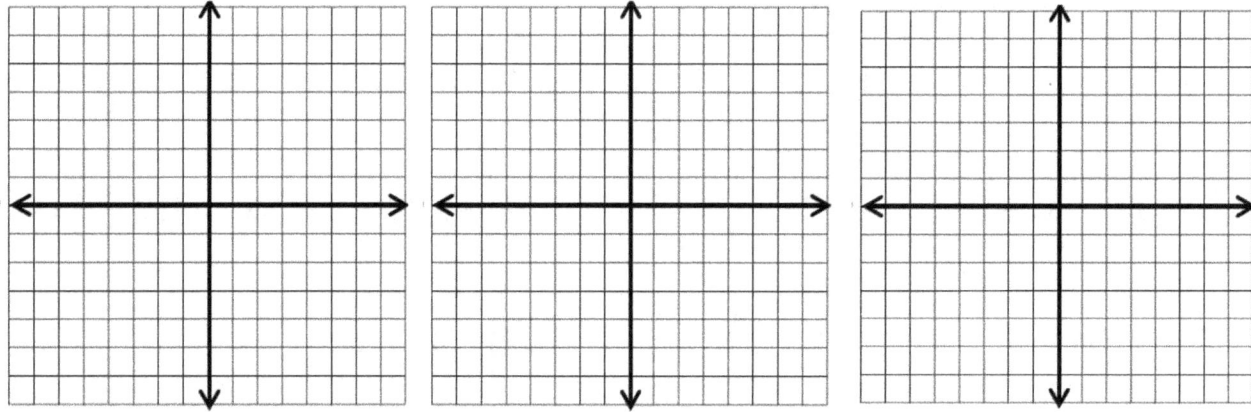

One Step Inequality

Solve each inequality.

1) $7x < 14$

2) $x + 7 \geq -8$

3) $x - 1 \leq 9$

4) $-2x + 4 > -10$

5) $x + 18 \geq -6$

6) $x + 9 \geq 5$

7) $x - \frac{1}{3} \leq 5$

8) $-7x < 42$

9) $-x + 8 > -3$

10) $\frac{x}{3} + 3 > -9$

11) $-x + 8 > -4$

12) $x - 14 \leq 18$

13) $-x - 5 \leq -7$

14) $x + 26 \geq -13$

15) $x + \frac{1}{3} \geq -\frac{2}{3}$

16) $x + 6 \geq -14$

17) $x - 42 \leq -48$

18) $x - 5 \leq 4$

19) $-x + 5 > -6$

20) $x + 6 \geq -12$

21) $8x + 6 \leq 22$

22) $4x - 3 \geq 9$

23) $3x - 5 < 22$

24) $6x - 8 \leq 40$

Two Steps Inequality

Solve each inequality

1) $2x - 3 \leq 7$

2) $3x - 4 \leq 8$

3) $\frac{-1}{4}x + \frac{x}{2} \leq \frac{1}{8}$

4) $5x + 10 \geq 30$

5) $4x - 7 \geq 9$

6) $3x - 5 \leq 16$

7) $8x - 2 \leq 14$

8) $9x + 5 \leq 23$

9) $2x + 10 > 32$

10) $\frac{x}{8} + 2 \leq 4$

11) $3x + 4 \geq 37$

12) $3x - 8 < 10$

13) $6 \geq \frac{x+7}{2}$

14) $3x + 9 < 48$

15) $\frac{4+x}{5} \geq 3$

16) $16 + 4x < 36$

17) $16 > 6x - 8$

18) $5 + \frac{x}{3} < 6$

19) $-4 + 4x > 24$

20) $5 + \frac{x}{7} < 3$

Multi Steps Inequality

Solve the inequalities.

1) $4x - 6 < 5x - 9$

2) $\frac{4x+5}{3} \leq x$

3) $7x - 5 > 3x + 15$

4) $-3x > -6x + 4$

5) $3 + \frac{x}{2} < \frac{x}{4}$

6) $\frac{4x-6}{8} > x$

7) $4x - 20 + 4 > 6x - 8$

8) $x - 8 > 11 + 3(x + 5)$

9) $\frac{x}{3} + 2 > x$

10) $-7x + 8 \geq -6(4x - 8) - 8x$

11) $7x - 4 \leq 8x + 9$

12) $\frac{2x-7}{5} > 2$

13) $8(x + 2) < 6x + 10$

14) $-8x + 12 \leq 4(x - 9)$

15) $\frac{5x-6}{3} > 3x + 2$

16) $2(x - 8) + 10 \geq 4x - 2$

17) $\frac{-5x+7}{6} > 5x$

18) $-3x - 4 > -7x$

19) $\frac{1}{4}x - 12 > \frac{1}{8}x - 19$

20) $-4(x - 9) \leq 5x$

Systems of Equations

Calculate each system of equations.

1) $-6x + 7y = 8$
   $x + 4y = 9$
   $x = \underline{\quad}$
   $y = \underline{\quad}$

2) $-4x + 12y = 12$
   $14x - 16y = 10$
   $x = \underline{\quad}$
   $y = \underline{\quad}$

3) $y = -9$
   $2x - 5y = 12$
   $x = \underline{\quad}$
   $y = \underline{\quad}$

4) $4y = -4x + 20$
   $8x - 2y = -12$
   $x = \underline{\quad}$
   $y = \underline{\quad}$

5) $10x - 9y = -13$
   $-5x + 3y = 11$
   $x = \underline{\quad}$
   $y = \underline{\quad}$

6) $-6x - 8y = 10$
   $4x - 8y = 20$
   $x = \underline{\quad}$
   $y = \underline{\quad}$

7) $5x - 14y = -23$
   $-6x + 7y = 8$
   $x = \underline{\quad}$
   $y = \underline{\quad}$

8) $-4x + 3y = 3$
   $-x + 2y = 5$
   $x = \underline{\quad}$
   $y = \underline{\quad}$

9) $-4x + 5y = 15$
   $-3x + 4y = -10$
   $x = \underline{\quad}$
   $y = \underline{\quad}$

10) $-6x - 6y = -21$
    $-6x + 6y = -66$
    $x = \underline{\quad}$
    $y = \underline{\quad}$

11) $12x - 21y = 6$
    $-6x - 3y = -12$
    $x = \underline{\quad}$
    $y = \underline{\quad}$

12) $-4x - 4y = -14$
    $4x - 4y = 44$
    $x = \underline{\quad}$
    $y = \underline{\quad}$

13) $4x + 5y = 3$
    $3x - y = 6$
    $x = \underline{\quad}$
    $y = \underline{\quad}$

14) $3x - 2y = 2$
    $10x - 10y = 20$
    $x = \underline{\quad}$
    $y = \underline{\quad}$

15) $5x + 8y = 14$
    $-3x - 2y = -3$
    $x = \underline{\quad}$
    $y = \underline{\quad}$

16) $8x + 5y = 4$
    $-3x - 4y = 15$
    $x = \underline{\quad}$
    $y = \underline{\quad}$

WWW.MathNotion.com

THEA Math Practice Book

Systems of Equations Word Problems

Find the answer for each word problem.

1) Tickets to a movie cost $6 for adults and $4 for students. A group of friends purchased 9 tickets for $50.00. How many adults ticket did they buy? ____

2) At a store, Eva bought two shirts and five hats for $77.00. Nicole bought three same shirts and four same hats for $84.00. What is the price of each shirt? ____

3) A farmhouse shelters 10 animals, some are pigs, and some are ducks. Altogether there are 36 legs. How many pigs are there? ____

4) A class of 85 students went on a field trip. They took 24 vehicles, some cars and some buses. If each car holds 3 students and each bus hold 16 students, how many buses did they take? ____

5) A theater is selling tickets for a performance. Mr. Smith purchased 8 senior tickets and 10 child tickets for $248 for his friends and family. Mr. Jackson purchased 4 senior tickets and 6 child tickets for $132. What is the price of a senior ticket? $____

6) The difference of two numbers is 15. Their sum is 33. What is the bigger number? $____

7) The sum of the digits of a certain two-digit number is 7. Reversing its digits increase the number by 9. What is the number? ____

8) The difference of two numbers is 11. Their sum is 25. What are the numbers? _____

9) The length of a rectangle is 5 meters greater than 2 times the width. The perimeter of rectangle is 28 meters. What is the length of the rectangle? _____

10) Jim has 23 nickels and dimes totaling $2.40. How many nickels does he have? ____

WWW.MathNotion.com

Finding Distance of Two Points

Find the distance between each pair of points.

1) $(2, 1), (-1, -3)$
2) $(-4, -2), (4, 4)$
3) $(-3, 0), (15, 24)$
4) $(-4, -1), (1, 11)$
5) $(3, -2), (-6, -14)$
6) $(-6, 0), (-2, 3)$
7) $(3, 2), (11, 17)$
8) $(-6, -10), (6, -1)$
9) $(5, 9), (-11, -3)$
10) $(6, -2), (2, -6)$
11) $(3, 0), (18, 36)$
12) $(8, 4), (3, -8)$
13) $(4, 2), (-5, -10)$
14) $(-8, 10), (4, 40)$
15) $(8, 4), (-10, -20)$
16) $(-8, -2), (16, 8)$
17) $(3, 5), (-5, -10)$
18) $(-10, 20), (35, 45)$

Find the midpoint of the line segment with the given endpoints.

1) $(-2, -2), (4, 2)$
2) $(10, 4), (-2, 4)$
3) $(12, -2), (4, 10)$
4) $(-6, -5), (2, 1)$
5) $(3, -2), (5, -2)$
6) $(-10, -4), (6, -2)$
7) $(4, 1), (-4, 9)$
8) $(-5, 6), (-5, 2)$
9) $(-8, 8), (4, -2)$
10) $(1, 7), (5, -1)$
11) $(-9, 5), (5, 3)$
12) $(7, 10), (-3, -6)$
13) $(-8, 14), (-8, 2)$
14) $(16, 7), (6, -3)$
15) $(5, 6), (-3, 4)$
16) $(-9, -1), (-5, 7)$
17) $(17, 9), (5, 11)$
18) $(-8, -11), (18, -1)$

Answer key Chapter 4

Distributive and Simplifying Expressions

1) $6x - 6$
2) $4 + 5x$
3) $6x - 8$
4) $-2x^2 - 6x$
5) $5x^2 - 2x$
6) $2x + 15y$
7) $11x - 6y$
8) $6x - 1$
9) $-8x + 21$
10) $-2x - 5$
11) $2x - 2y$
12) $-12.5x^3$
13) $4x^2 - 4$
14) $14x^2 + 12$
15) $-8x - 18$
16) $-4x^2 - 5x$
17) $2x - 54$
18) $60x - 20$
19) $-6x - 18$
20) $7x$
21) $-2x^2 + 8x$
22) $5x + y + 5xy$
23) $-3x - 12$
24) $-3x + 3$
25) $3x + 4y - 4$
26) $-6x^2 - 2$
27) $9x + 9y$
28) $-4x - 3$

Factoring Expressions

1) $3(4x - 2)$
2) $5(x - 3)$
3) $3(x - 5)$
4) $7(b - 4)$
5) $4a(a - 6)$
6) $2y(x - 5)$
7) $5x(xy + 3)$
8) $a(a - 8 + 7b)$
9) $2a(a + b)$
10) $4(x + 5)$
11) $12x(2 - 3y)$
12) $2(4x - 3)$
13) $\frac{1}{4}(x - 3y)$
14) $7x(y - \frac{2}{3})$
15) $3(ab + 3c)$
16) $\frac{1}{3}(x - 4)$
17) $5x(2 - 3y)$
18) $x(x + 8)$
19) $4(x^2 - 3y)$
20) $x(4x^2 + 3y + x)$
21) $7(3x - 2)$
22) $20(b - 3c + d)$
23) $8a(3b - c)$
24) $(x - y)(a - 3)$
25) $(3x + 4)(a + 3)$
26) $x(x - 10)$
27) $9x^2(x - 2)$
28) $5x(x - 14y)$

Evaluate One Variable Expressions

1) 15
2) -15
3) -33
4) 0
5) 10
6) -4
7) 11
8) 6
9) -21
10) 6
11) 58
12) 26
13) 4
14) 5
15) 17
16) 16
17) 36
18) 6
19) 4
20) -33
21) 13
22) 4
23) -5
24) -14

THEA Math Practice Book

25) 9   26) 28   27) 26   28) 27

Evaluate Two Variable Expressions

1) 13
2) 14
3) 50
4) 1
5) −16
6) 35
7) 16
8) 7
9) 27
10) 3
11) 18
12) 20
13) 17
14) 78
15) 10
16) 210
17) −7
18) 22
19) 8
20) 70
21) 25
22) 80
23) 36
24) 10
25) −3
26) 270
27) 5
28) $\frac{10}{7}$

Graphing Lines Using Line Equation

1) $y = 2x - 5$

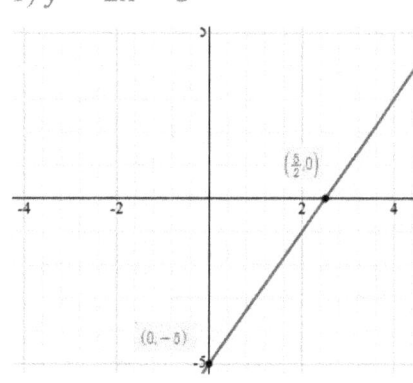

2) $y = -2x + 3$

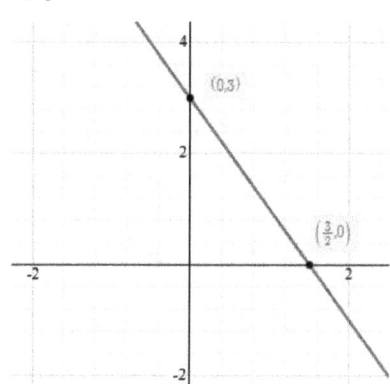

3) $x - y = 0$

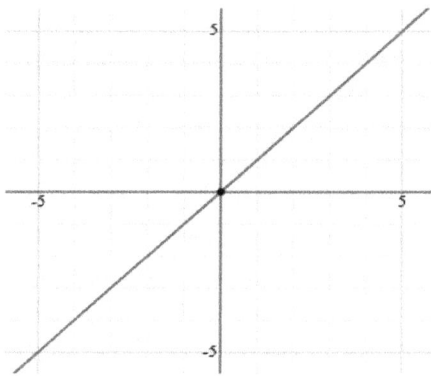

4) $x + y = 3$

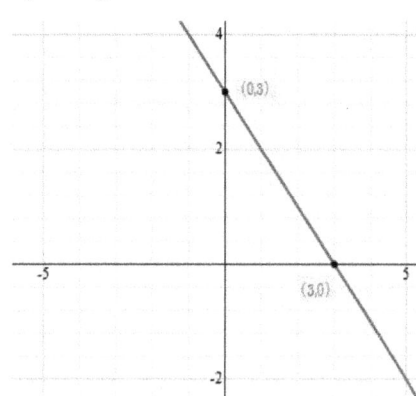

5) $5x + 3y = -2$

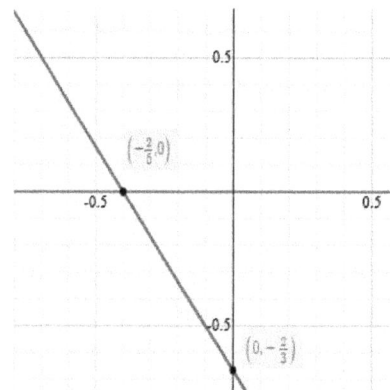

6) $y - 3x + 2 = 0$

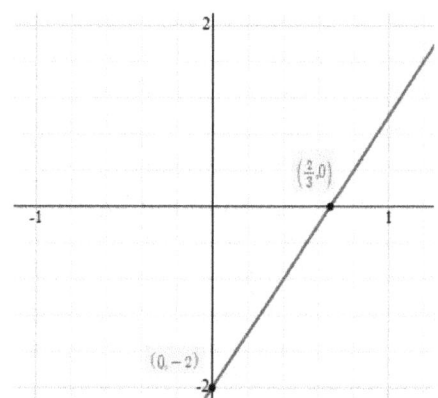

One Step Equations

1) $x = 56$
2) $x = -8$
3) $x = 9$
4) $x = -8$
5) $x = -14$
6) $x = -13$
7) x = 75
8) x = -45
9) x = 42
10) x = 1
11) x = 8
12) x = -13
13) x = -13
14) x = -2
15) x = 7
16) x = -10
17) $x = 4$
18) x = 6
19) x = 33
20) x = 5
21) x = 21
22) x = 32
23) $x = -3$
24)
25)
26)

Two Steps Equations

1) $x = 4$
2) $x = -6$
3) $x = 2$
4) $x = -5$
5) $x = -3$
6) $x = 2$
7) $x = -4$
8) $x = -1$
9) $x = 2$
10) $x = 5$
11) $x = -25$
12) $x = -7$
13) $x = 36$
14) $x = -1$
15) $x = -41$
16) $x = -1$
17) $x = -4$
18) $x = -30$
19) $x = 1$
20) $x = -20$
21) $x = -47$
22) $x = 3$
23) $x = 30$
24) $x = 0$

Multi Steps Equations

1) $x = 1$
2) $x = 2$
3) $x = 3$
4) $x = 2$
5) $x = -3$
6) $x = 13$
7) $x = 8$
8) $x = 2$
9) $x = -4$
10) $x = -24$
11) $x = -0.1$
12) $x = 7$
13) $x = 8$
14) $x = 3$
15) $x = -4$
16) $x = 5$
17) $x = 5$
18) $x = -3/5$
19) $x = -3$
20) $x = -3$
21) $x = -3$
22) $x = 2$
23) $x = 4$
24) $x = 3$

THEA Math Practice Book

Graphing Linear Inequalities

1) $y > 2x - 3$
2) $y < x + 3$
3) $y \leq -3x - 8$

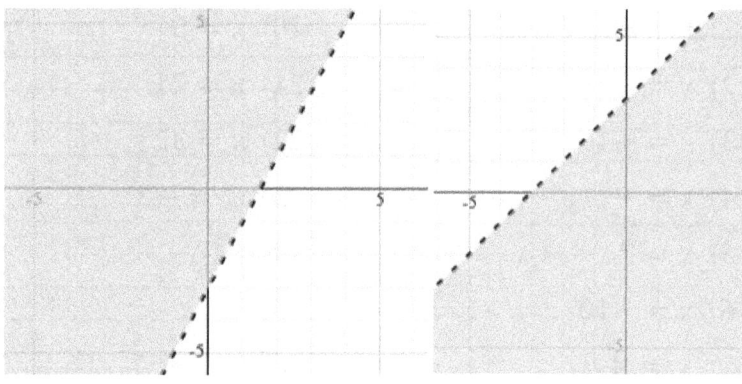
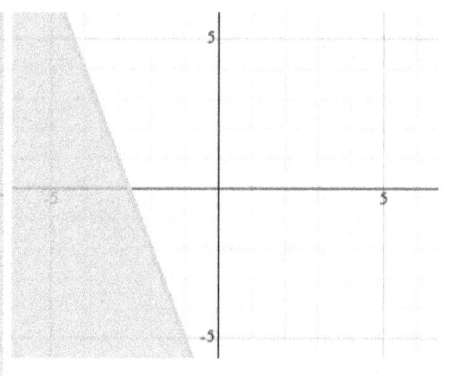

4) $3y \geq 6 + 3x$
5) $-3y < x - 12$
6) $2y \geq -8x + 4$

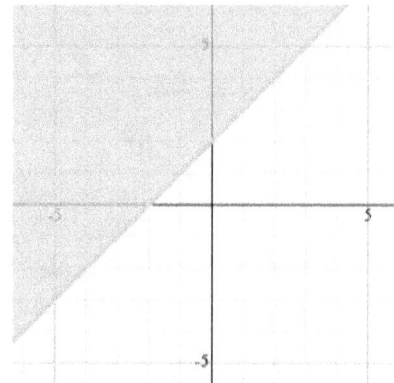
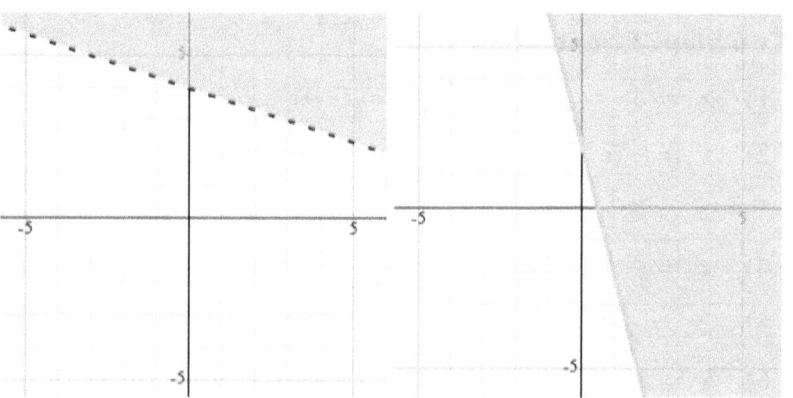

One Step Inequality

1) $x < 2$
2) $x \geq -15$
3) $x \leq 10$
4) $x < 7$
5) $x \geq -24$
6) $x \geq -4$
7) $x \leq \frac{16}{3}$
8) $x > -6$
9) $x < 11$
10) $x > -36$
11) $x < 12$
12) $x \leq 32$
13) $x \geq 2$
14) $x \geq -39$
15) $x \geq -1$
16) $x \geq -20$
17) $x \leq -6$
18) $x \leq 9$
19) $x < 11$
20) $x \geq -18$
21) $x \leq 2$
22) $x \geq 3$
23) $x < 9$
24) $x \leq 8$

Two Steps Inequality

1) $x \leq 5$
2) $x \leq 4$
3) $x \leq 0.5$
4) $x \geq 4$
5) $x \geq 4$
6) $x \leq 7$
7) $x \leq 2$
8) $x \leq 2$
9) $x > 11$

10) $x \leq 16$
11) $x \geq 11$
12) $x < 6$
13) $x \leq 5$

14) $x < 13$
15) $x \geq 11$
16) $x < 5$
17) $x < 4$

18) $x < 3$
19) $x > 7$
20) $x < -14$

Multi Steps Inequality

1) $x > 3$
2) $x \leq -5$
3) $x > 5$
4) $x > \frac{4}{3}$
5) $x < -12$
6) $x < -1.5$
7) $x < -4$

8) $x < -17$
9) $x < 3$
10) $x \geq 1.6$
11) $x \geq -13$
12) $x > 8.5$
13) $x < -3$
14) $x \geq 4$

15) $x < -3$
16) $x \leq -2$
17) $x < \frac{1}{5}$
18) $x > 1$
19) $x > -56$
20) $x \geq 4$

Systems of Equations

1) $x = 1, y = 2$
2) $x = 3, y = -2$
3) $x = -\frac{33}{2}$
4) $x = -\frac{1}{5}, y = \frac{26}{5}$
5) $x = -4, y = -3$
6) $x = 1, y = -2$

7) $x = 1, y = 2$
8) $x = \frac{9}{5}, y = \frac{17}{5}$
9) $x = -110, y = -85$
10) $x = -\frac{15}{4}, y = \frac{29}{4}$
11) $x = \frac{5}{3}, y = \frac{2}{3}$
12) $x = -\frac{15}{4}, y = \frac{29}{4}$

13) $x = \frac{33}{19}, y = -\frac{15}{19}$
14) $x = -2, y = -4$
15) $x = -\frac{2}{7}, y = \frac{27}{14}$
16) $x = \frac{91}{17}, y = -\frac{132}{17}$

Systems of Equations Word Problems

1) 7
2) $16
3) 8
4) 1

5) $21
6) 24
7) 43
8) 18, 7

9) 11 meters
10) 18

Finding Distance of Two Points

1) 5
2) 10
3) 30
4) 13
5) 15

6) 5
7) 17
8) 15
9) 20
10) $4\sqrt{2}$

11) 39
12) 13
13) 15
14) $6\sqrt{29}$
15) 30

16) 26   17) 17   18) $5\sqrt{106}$

Finding Midpoint

1) $(1, 0)$ 7) $(0, 5)$ 13) $(-8, 8)$
2) $(4, 4)$ 8) $(-5, 4)$ 14) $(11, 2)$
3) $(8, 4)$ 9) $(-2, 3)$ 15) $(1, 5)$
4) $(-2, -2)$ 10) $(3, 3)$ 16) $(-7, 3)$
5) $(4, -2)$ 11) $(-2, 4)$ 17) $(11, 10)$
6) $(-2, -3)$ 12) $(2, 2)$ 18) $(5, -6)$

Chapter 5: Exponent and Radicals

Positive Exponents

Simplify. Your answer should contain only positive exponents.

1) $3^4 =$

2) $2^5 =$

3) $\frac{3x^6y}{xy} =$

4) $(12x^2x)^3 =$

5) $(x^2)^4 =$

6) $(\frac{1}{4})^3 =$

7) $0^8 =$

8) $6 \times 6 \times 6 =$

9) $3 \times 3 \times 3 \times 3 \times 3 =$

10) $(4x^4y)^2 =$

11) $9^3 =$

12) $(5x^3y^2)^2 =$

13) $5 \times 10^4 =$

14) $0.6 \times 0.6 \times 0.6 =$

15) $\frac{1}{3} \times \frac{1}{3} \times \frac{1}{3} =$

16) $4^5 =$

17) $(5x^8y^2)^3 =$

18) $7^3 =$

19) $y \times y \times y \times y =$

20) $8 \times 8 \times 8 \times 8 =$

21) $(2x^4y^2z)^3 =$

22) $8^0 =$

23) $(11x^4y^{-1})^4 =$

24) $(2x^2y^4)^5 =$

Negative Exponents

Simplify. Leave no negative exponents.

1) $2^{-4} =$

2) $9^{-2} =$

3) $\left(\frac{1}{3}\right)^{-2} =$

4) $8^{-3} =$

5) $1^{150} =$

6) $6^{-3} =$

7) $\left(\frac{1}{2}\right)^{-6} =$

8) $-8y^{-4} =$

9) $\left(\frac{1}{y^{-5}}\right)^{-3} =$

10) $x^{-\frac{4}{5}} =$

11) $\frac{1}{7^{-6}} =$

12) $3^{-5} =$

13) $5^{-2} =$

14) $13^{-2} =$

15) $30^{-2} =$

16) $x^{-8} =$

17) $(x^2)^{-4} =$

18) $x^{-2} \times x^{-2} \times x^{-2} \times x^{-2} =$

19) $\frac{1}{3} \times \frac{1}{3} =$

20) $100^{-2} =$

21) $100z^{-3} =$

22) $3^{-4} =$

23) $\left(-\frac{1}{11}\right)^2 =$

24) $14^0 =$

25) $\left(\frac{1}{x}\right)^{-18} =$

26) $15^{-2} =$

Add and subtract Exponents

Solve each problem.

1) $4^2 + 5^3 =$

2) $x^8 + x^8 =$

3) $5b^3 - 4b^3 =$

4) $6 + 5^2 =$

5) $9 - 6^2 =$

6) $12 + 3^2 =$

7) $5x^2 + 8x^2 =$

8) $9^2 + 2^6 =$

9) $3^6 - 4^3 =$

10) $8^2 - 10^0 =$

11) $7^2 - 4^2 =$

12) $9^2 + 3^4 =$

13) $12^2 - 5^2 =$

14) $7^2 + 7^2 =$

15) $6^3 - 4^3 =$

16) $1^{24} + 1^{28} =$

17) $4^3 - 2^3 =$

18) $5^4 - 5^2 =$

19) $7^2 - 4^2 =$

20) $5^2 + 8^2 =$

21) $4^2 + 3^4 =$

22) $18 + 2^4 =$

23) $7x^8 + 5x^8 =$

24) $9^0 + 8^2 =$

25) $5^2 + 5^2 =$

26) $10^3 + 2^2 =$

27) $(\frac{1}{3})^2 + (\frac{1}{3})^2 =$

28) $8^2 + 2^2 =$

Exponent multiplication

Simplify each of the following

1) $2^5 \times 2^3 =$

2) $7^2 \times 8^0 =$

3) $9^1 \times 4^2 =$

4) $a^{-5} \times a^{-5} =$

5) $y^{-3} \times y^{-3} \times y^{-3} =$

6) $4^5 \times 5^7 \times 4^{-4} \times 5^{-6} =$

7) $6x^4y^3 \times 4x^3y^2 =$

8) $(x^3)^5 =$

9) $(x^4y^6)^5 \times (x^4y^5)^{-5} =$

10) $8^4 \times 8^2 =$

11) $a^{4b} \times a^0 =$

12) $4^2 \times 4^2 =$

13) $a^{3m} \times a^{2n} =$

14) $2a^n \times 4b^n =$

15) $5^{-3} \times 4^{-3} =$

16) $6^{10} \times 3^{10} =$

17) $(7^6)^5 =$

18) $(\frac{1}{6})^2 \times (\frac{1}{6})^4 \times (\frac{1}{6})^5 =$

19) $(\frac{1}{9})^{52} \times 9^{52} =$

20) $(4m)^{\frac{4}{5}} \times (-2m)^{\frac{4}{5}} =$

21) $(x^4y)^{\frac{1}{4}} \times (xy^3)^{\frac{1}{4}} =$

22) $(2a^mb^n)^r =$

23) $(5x^3y^2)^3 =$

24) $(x^{\frac{1}{3}}y^2)^{\frac{-1}{3}} \times (x^4y^6)^0 =$

25) $7^6 \times 7^5 =$

26) $28^{\frac{1}{6}} \times 28^{\frac{1}{3}} =$

27) $9^5 \times 3^5 =$

28) $(x^{12})^0 =$

Exponent division

Simplify. Your answer should contain only positive exponents.

1) $\dfrac{5^4}{5} =$

2) $\dfrac{38x^4}{x} =$

3) $\dfrac{a^m}{a^{2n}} =$

4) $\dfrac{3x^{-6}}{15x^{-4}} =$

5) $\dfrac{63x^9}{7x^4} =$

6) $\dfrac{17x^7}{5x^8} =$

7) $\dfrac{36x^8}{12y^3} =$

8) $\dfrac{45xy^6}{x^4y^2} =$

9) $\dfrac{3x^9}{8x} =$

10) $\dfrac{45x^7y^9}{5x^8} =$

11) $\dfrac{12x^4}{20x^9y^{12}} =$

12) $\dfrac{8yx^7}{40yx^{10}} =$

13) $\dfrac{21x^3y^2}{3x^2y^3} =$

14) $\dfrac{x^{4.75}}{x^{0.75}} =$

15) $\dfrac{9x^4y}{18xy^3} =$

16) $\dfrac{34b^3r^8}{17a^2b^5} =$

17) $\dfrac{30x^7}{15x^9} =$

18) $\dfrac{44x^5}{11x^8} =$

19) $\dfrac{6^5}{6^3} =$

20) $\dfrac{x}{x^{10}} =$

21) $\dfrac{13^7}{13^4} =$

22) $\dfrac{3xy^5}{12y^3} =$

23) $\dfrac{13x^6y}{169xy^3} =$

24) $\dfrac{48x^5}{8y^9} =$

Scientific Notation

Write each number in scientific notation.

1) 9,500,000 =

2) 800 =

3) 0.000007 =

4) 387,000 =

5) 0.00139 =

6) 0.85 =

7) 0.000093 =

8) 20,000,000 =

9) 28,000,000 =

10) 230,000,000 =

11) 0.000049 =

12) 0.00002 =

13) 0.00027 =

14) 70,000 =

15) 2,870 =

16) 190,000 =

17) 0.0223 =

18) 0.7 =

19) 0.082 =

20) 310,000 =

21) 48,000 =

22) 0.000098 =

23) 0.035 =

24) 1,778 =

25) 58,781 =

26) 24,500 =

27) 33,021 =

28) 8,100,000 =

Square Roots

Find the square root of each number.

1) $\sqrt{64} =$

2) $\sqrt{0} =$

3) $\sqrt{324} =$

4) $\sqrt{484} =$

5) $\sqrt{1,600} =$

6) $\sqrt{529} =$

7) $\sqrt{0.01} =$

8) $\sqrt{10,000} =$

9) $\sqrt{0.16} =$

10) $\sqrt{0.36} =$

11) $\sqrt{0.25} =$

12) $\sqrt{1.21} =$

13) $\sqrt{784} =$

14) $\sqrt{576} =$

15) $\sqrt{676} =$

16) $\sqrt{961} =$

17) $\sqrt{1,681} =$

18) $\sqrt{0.81} =$

19) $\sqrt{0.49} =$

20) $\sqrt{0.64} =$

21) $\sqrt{1,089} =$

22) $\sqrt{2,500} =$

23) $\sqrt{8,100} =$

24) $\sqrt{12,100} =$

25) $\sqrt{2.25} =$

26) $\sqrt{1.69} =$

27) $\sqrt{1.44} =$

28) $\sqrt{0.04} =$

Simplify Square Roots

Simplify the following.

1) $\sqrt{54} =$

2) $\sqrt{108} =$

3) $\sqrt{12} =$

4) $\sqrt{99} =$

5) $\sqrt{200} =$

6) $\sqrt{45} =$

7) $8\sqrt{50} =$

8) $3\sqrt{300} =$

9) $\sqrt{24} =$

10) $2\sqrt{18} =$

11) $4\sqrt{3} + 7\sqrt{3} =$

12) $\frac{11}{4+\sqrt{5}} =$

13) $\sqrt{48} =$

14) $\frac{4}{3-\sqrt{5}} =$

15) $\sqrt{18} \times \sqrt{2} =$

16) $\frac{\sqrt{300}}{\sqrt{3}} =$

17) $\frac{\sqrt{90}}{\sqrt{18 \times 5}} =$

18) $\sqrt{80y^6} =$

19) $6\sqrt{81a} =$

20) $\sqrt{41+8} + \sqrt{9} =$

21) $\sqrt{72} =$

22) $\sqrt{432} =$

23) $\sqrt{112} =$

24) $\sqrt{128} =$

25) $\sqrt{768} =$

26) $\sqrt{96} =$

Answer key Chapter 5

Positive Exponents

1) 81
2) 32
3) $3x^5$
4) $1,728x^9$
5) x^8
6) $\frac{1}{64}$
7) 0
8) 6^3
9) 3^5
10) $16x^8y^2$
11) 729
12) $25x^6y^4$
13) 50,000
14) 0.6^3
15) $(\frac{1}{3})^3$
16) 1,024
17) $125x^{24}y^6$
18) 343
19) y^4
20) 8^4
21) $8x^{12}y^6z^3$
22) 1
23) $\frac{121x^8}{y^4}$
24) $32x^{10}y^{20}$

Negative Exponents

1) $\frac{1}{16}$
2) $\frac{1}{81}$
3) 9
4) $\frac{1}{512}$
5) 1
6) $\frac{1}{216}$
7) 64
8) $\frac{-8}{y^4}$
9) y^{15}
10) $\frac{1}{x^{\frac{4}{5}}}$
11) 7^6
12) $\frac{1}{243}$
13) $\frac{1}{25}$
14) $\frac{1}{169}$
15) $\frac{1}{900}$
16) $\frac{1}{x^8}$
17) $\frac{1}{x^8}$
18) $\frac{1}{x^8}$
19) $\frac{1}{3^2}$
20) $\frac{1}{10,000}$
21) $\frac{100}{z^3}$
22) $\frac{1}{81}$
23) $\frac{1}{121}$
24) 1
25) x^{18}
26) $\frac{1}{225}$

Add and subtract Exponents

1) 141
2) $2x^8$
3) b^3
4) 31
5) −27
6) 21
7) $13x^2$
8) 145
9) 665
10) 63
11) 33
12) 162
13) 119
14) 98
15) 152

THEA Math Practice Book

16) 1
17) 56
18) 600
19) 33
20) 89

21) 97
22) 34
23) $12x^8$
24) 65
25) 50

26) 1,004
27) $\frac{2}{9}$
28) 68

Exponent multiplication

1) 2^8
2) 49
3) 144
4) a^{-10}
5) y^{-9}
6) 20
7) $24x^7y^5$
8) x^{15}
9) y^5
10) 8^6

11) a^{4b}
12) 4^4
13) a^{3m+2n}
14) $8(ab)^n$
15) 20^{-3}
16) 18^{10}
17) 7^{30}
18) $\left(\frac{1}{6}\right)^{11}$
19) 1

20) $(-8m^2)^{\frac{4}{5}}$
21) $x^{\frac{5}{4}}y$
22) $2^r a^{mr} b^{nr}$
23) $125x^9y^6$
24) $x^{\frac{5}{4}}y$
25) 7^{11}
26) $28^{\frac{1}{2}}$
27) $27^5 = 3^{15}$
28) 1

Exponent division

1) 5^3
2) $38x^3$
3) a^{m-2n}
4) $\frac{1}{5x^2}$
5) $9x^5$
6) $\frac{17}{5x}$
7) $\frac{3x^8}{y^3}$
8) $\frac{45y^4}{x^3}$

9) $\frac{3x^8}{8}$
10) $\frac{9y^9}{x}$
11) $\frac{3}{5x^5y^{12}}$
12) $\frac{1}{5x^3}$
13) $\frac{7x}{y}$
14) x^4
15) $\frac{x^3}{2y^2}$
16) $\frac{2r^8}{a^2b^2}$

17) $\frac{2}{x^2}$
18) $\frac{4}{x^3}$
19) 6^2
20) $\frac{1}{x^9}$
21) 13^3
22) $\frac{1}{4}xy^2$
23) $\frac{x^5}{13y^2}$
24) $\frac{6x^5}{y^9}$

Scientific Notation

1) 9.5×10^6
2) 8×10^2

3) 7×10^{-6}
4) 3.87×10^5

5) 1.39×10^{-3}
6) 8.5×10^{-1}

7) 9.3×10^{-5}
8) 2×10^7
9) 2.8×10^7
10) 2.3×10^8
11) 4.9×10^{-5}
12) 2×10^{-5}
13) 2.7×10^{-4}
14) 7×10^4
15) 2.87×10^3
16) 1.9×10^5
17) 2.23×10^{-2}
18) 7×10^{-1}
19) 8.2×10^{-2}
20) 3.1×10^5
21) 4.8×10^4
22) 9.8×10^{-5}
23) 3.5×10^{-2}
24) 1.778×10^3
25) 5.8781×10^4
26) 2.45×10^4
27) 3.3021×10^4
28) 8.1×10^6

Square Roots

1) 8
2) 0
3) 18
4) 22
5) 40
6) 23
7) 0.1
8) 100
9) 0.4
10) 0.6
11) 0.5
12) 1.1
13) 28
14) 24
15) 26
16) 31
17) 41
18) 0.9
19) 0.7
20) 0.8
21) 33
22) 50
23) 90
24) 110
25) 1.5
26) 1.3
27) 1.2
28) 0.2

Simplify Square Roots

1) $3\sqrt{6}$
2) $6\sqrt{3}$
3) $2\sqrt{3}$
4) $3\sqrt{11}$
5) $10\sqrt{2}$
6) $3\sqrt{5}$
7) $40\sqrt{2}$
8) $30\sqrt{3}$
9) $2\sqrt{6}$
10) $6\sqrt{2}$
11) $11\sqrt{3}$
12) $4 - \sqrt{5}$
13) $4\sqrt{3}$
14) $3 + \sqrt{5}$
15) 6
16) 10
17) 1
18) $4y^3\sqrt{5}$
19) $54\sqrt{a}$
20) 10
21) $6\sqrt{2}$
22) $12\sqrt{3}$
23) $4\sqrt{7}$
24) $8\sqrt{2}$
25) $16\sqrt{3}$
26) $4\sqrt{6}$

Chapter 6: Ratio, Proportion and Percent

Proportions

Find a missing number in a proportion.

1) $\frac{4}{7} = \frac{12}{a}$

2) $\frac{a}{9} = \frac{20}{45}$

3) $\frac{12}{60} = \frac{a}{5}$

4) $\frac{16}{a} = \frac{96}{36}$

5) $\frac{4}{a} = \frac{16}{75}$

6) $\frac{\sqrt{9}}{4} = \frac{a}{32}$

7) $\frac{2}{4} = \frac{18}{a}$

8) $\frac{7}{14} = \frac{a}{35}$

9) $\frac{7}{a} = \frac{4.2}{6}$

10) $\frac{2}{12} = \frac{8}{a}$

11) $\frac{10}{6} = \frac{5}{a}$

12) $\frac{16}{a} = \frac{4}{19}$

13) $\frac{4}{11} = \frac{a}{12}$

14) $\frac{\sqrt{36}}{3} = \frac{48}{a}$

15) $\frac{6}{a} = \frac{6.6}{39.6}$

16) $\frac{60}{140} = \frac{a}{280}$

17) $\frac{42}{200} = \frac{a}{68}$

18) $\frac{26}{104} = \frac{a}{4}$

19) $\frac{10}{16} = \frac{2}{a}$

20) $\frac{9}{7} = \frac{27}{a}$

Reduce Ratio

Reduce each ratio to the simplest form.

1) 5: 20 =

2) 6: 36 =

3) 63: 35 =

4) 24: 20 =

5) 12: 120 =

6) 16: 2 =

7) 70: 350 =

8) 4: 144 =

9) 25: 75 =

10) 4.8: 5.6 =

11) 110: 330 =

12) 3: 5 =

13) 120: 200 =

14) 30: 45 =

15) 34: 68 =

16) 32: 8 =

17) 140: 35 =

18) 20: 200 =

19) 126: 84 =

20) 156: 198 =

21) 40: 80 =

22) 42: 49 =

23) 5: 75 =

24) 18: 108 =

Percent

Find the Percent of Numbers.

1) 30% of 42 =

2) 28% of 15 =

3) 15% of 14 =

4) 24% of 70 =

5) 8% of 80 =

6) 35% of 12 =

7) 18% of 5 =

8) 12% of 46 =

9) 40% of 62 =

10) 4.5% of 50 =

11) 85% of 18 =

12) 60% of 50 =

13) 18% of 180 =

14) 2% of 240 =

15) 75% of 0 =

16) 80% of 120 =

17) 36% of 45 =

18) 10% of 70 =

19) 8% of 13 =

20) 4% of 8 =

21) 30% of 44 =

22) 80% of 17 =

23) 22% of 35 =

24) 8% of 150 =

25) 40% of 270 =

26) 2% of 5 =

27) 9% of 320 =

28) 10% of 26 =

Discount, Tax and Tip

Find the selling price of each item.

1) Original price of a computer: $150
   Tax: 8%, Selling price: $_____

2) Original price of a laptop: $240
   Tax: 15%, Selling price: $_____

3) Original price of a sofa: $300
   Tax: 8%, Selling price: $_____

4) Original price of a car: $12,600
   Tax: 3.5%, Selling price: $_____

5) Original price of a Table: $500
   Tax: 12%, Selling price: $_____

6) Original price of a house: $280,000
   Tax: 1.5%, Selling price: $_____

7) Original price of a tablet: $460
   Discount: 30%, Selling price: $_____

8) Original price of a chair: $110
   Discount: 10%, Selling price: $_____

9) Original price of a book: $30
   Discount: 10% Selling price: $_____

10) Original price of a cellphone: 720
    Discount: 12% Selling price: $_____

11) Food bill: $27
    Tip: 15%   Price: $_____

12) Food bill: $45
    Tipp: 10%   Price: $_____

13) Food bill: $50
    Tip: 25%   Price: $_____

14) Food bill: $72
    Tipp: 18%   Price: $_____

Find the answer for each word problem.

15) Nicolas hired a moving company. The company charged $400 for its services, and Nicolas gives the movers a 25% tip. How much does Nicolas tip the movers? $_____

16) Mason has lunch at a restaurant and the cost of his meal is $80. Mason wants to leave a 8% tip. What is Mason's total bill including tip? $_____

Percent of Change

Find each percent of change.

1) From 300 to 600. ___%

2) From 45 ft to 225 ft. ___%

3) From $60 to $420. ___%

4) From 50 cm to 150 cm. ___%

5) From 10 to 30. ___%

6) From 60 to 108. ___%

7) From 120 to 180. ___%

8) From 400 to 600. ___%

9) From 85 to 119. ___%

10) From 100 to 175. ___%

Calculate each percent of change word problem.

11) Bob got a raise, and his hourly wage increased from $32 to $40. What is the percent increase? ____%

12) The price of a pair of shoes increases from $50 to $80. What is the percent increase? ___%

13) At a coffee shop, the price of a cup of coffee increased from $3.50 to $4.2. What is the percent increase in the cost of the coffee? _____%

14) 22cm are cut from a 88 cm board. What is the percent decrease in length? _%

15) In a class, the number of students has been increased from 112 to 168. What is the percent increase? _____%

16) The price of gasoline rose from $18.4 to $21.16 in one month. By what percent did the gas price rise? _____%

17) A shirt was originally priced at $42. It went on sale for $33.6. What was the percent that the shirt was discounted? _____%

Simple Interest

Determine the simple interest for these loans.

1) $140 at 18% for 5 years. $ _____
2) $1,800 at 6% for 2 years. $ _____
3) $900 at 25% for 4 years. $ _____
4) $9,200 at 1.5% for 8 months. $ ___
5) $600 at 5% for 7 months. $ _____

6) $40,000 at 8.5% for 3 years. $ ____
7) $7,400 at 8% for 5 years. $ _____
8) $500 at 2.5% for 2 years. $ _____
9) $600 at 6.5 % for 4 months. $ ____
10) $8,000 at 3.5% for 4 years. $ ___

Calculate each simple interest word problem.

11) A new car, valued at $18,000, depreciates at 5.5% per year. What is the value of the car one year after purchase? $_____

12) Sara puts $9,000 into an investment yielding 8% annual simple interest; she left the money in for two years. How much interest does Sara get at the end of those three years? $_____

13) A bank is offering 12.5% simple interest on a savings account. If you deposit $30,400, how much interest will you earn in four years? $_____

14) $1,200 interest is earned on a principal of $5,000 at a simple interest rate of 12% interest per year. For how many years was the principal invested? _____

15) In how many years will $1,800 yield an interest of $432 at 6% simple interest? _____

16) Jim invested $8,000 in a bond at a yearly rate of 2.5%. He earned $600 in interest. How long was the money invested? _____

Answer key Chapter 6

Proportions

1) $a = 21$
2) $a = 4$
3) $a = 1$
4) $a = 6$
5) $a = 18.75$
6) $a = 24$
7) $a = 36$
8) $a = 17.5$
9) $a = 10$
10) $a = 48$
11) $a = 3$
12) $a = 76$
13) $a = \frac{48}{11}$
14) $a = 24$
15) $a = 36$
16) $a = 120$
17) $a = 14.28$
18) $a = 1$
19) $a = 3.2$
20) $a = 21$

Reduce Ratio

1) 1: 4
2) 1: 6
3) 9: 5
4) 6: 5
5) 1: 10
6) 8: 1
7) 1: 5
8) 1: 36
9) 1: 3
10) 0.6: 0.7
11) 11: 33
12) 0.6: 1
13) 3: 5
14) 2: 3
15) 1: 2
16) 4: 1
17) 4: 1
18) 1: 10
19) 3: 2
20) 26: 33
21) 1: 2
22) 6: 7
23) 1: 15
24) 1: 6

Percent

1) 12.6
2) 4.2
3) 2.1
4) 16.8
5) 6.4
6) 4.2
7) 0.9
8) 5.52
9) 24.8
10) 2.25
11) 15.3
12) 30
13) 13.4
14) 4.8
15) 0
16) 96
17) 16.2
18) 7
19) 1.04
20) 0.32
21) 13.2
22) 13.6
23) 7.7
24) 12
25) 108
26) 0.1
27) 28.8
28) 2.6

Discount, Tax and Tip

1) $162.00
2) $276.00
3) $324.00
4) $13,041.00
5) $560.00
6) $284,200
7) $322.00
8) $99.00
9) $27.00
10) $633.60
11) $31.05
12) $49.50
13) $62.50
14) $84.96
15) $100.00
16) $86.40

Percent of Change

1) 100%
2) 400%
3) 600%
4) 300%
5) 200%
6) 80%
7) 50%
8) 50%
9) 40%
10) 75%
11) 25%
12) 60%
13) 20%
14) 25%
15) 50%
16) 15%
17) 20%

Simple Interest

1) $126.00
2) $216.00
3) $900.00
4) $92.00
5) $17.50
6) $10,200.00
7) $2,960.00
8) $25.00
9) $13.00
10) $1,120.00
11) $17,010.00
12) $2,160.00
13) $15,200.00
14) 2 years
15) 4 years
16) 3 years

Chapter 7: Monomials and Polynomials

Adding and Subtracting Monomial

Simplify each expression.

1) $3x^3 + 10x^3 =$

2) $5x^2 + 2x^2 =$

3) $\frac{1}{8}x^3 + \frac{4}{8}x^3 =$

4) $3\frac{1}{4}x^4 + 5\frac{3}{4}x^4 =$

5) $11x^7 - 5x^7 =$

6) $6.9x^3 - 2.9x^3 =$

7) $-x^{10} + x^{10} =$

8) $(x^4)^5 + (x^5)^4 =$

9) $3x^{-5} + 2x^{-5} =$

10) $15p^8 - (-5p^8) =$

11) $2x^2 - 3.8x^2 =$

12) $4\frac{1}{5}x^4 + 3\frac{2}{5}x^4 =$

13) $-4\frac{1}{8}x^{13} + 6\frac{1}{8}x^{13} =$

14) $\sqrt{81}p^6 + (-5p^6) =$

15) $(-1.8p^4) + (-3.2p^4) =$

16) $-1.2x^6 + 7.4x^6 =$

17) $x^8 + \frac{2}{5}x^8 =$

18) $12x^4 - x^4 =$

19) $-3.2x^2 - 6.9x^2 =$

20) $-6x^7 - 3x^7 =$

21) $32x^5 - 20x^5 =$

22) $-3x^{13} + 7x^{13} =$

23) $5x^{-10} - 16x^{-10} =$

24) $17x^{-7} - 7x^{-7} =$

Multiplying and Dividing Monomial

Simplify.

1) $5xy^3 \times 2x^4 =$

2) $7xy \times 4x^3 =$

3) $5xy^3 \times (-6x^3y^4) =$

4) $5x^6y^{10} \times x^3y^2 =$

5) $12x^2 \times (-5x^4) =$

6) $-5x^2y^4z \times 3x^3y^3z^4 =$

7) $-7 \times (-11x^{14}y^{16}) =$

8) $3x^2y^4 \times (-12x^2y^5) =$

9) $6x^2 \times (-8x) =$

10) $-8x^3y^8 \times 3x^2y =$

11) $22x^{-4}y^6 \times (-x^{-7}y^{-3}) =$

12) $6x^{10}y^3z \times 5xy^{-3}z =$

13) $\frac{20x^{11}y^4}{10x^5y^2} =$

14) $(6y^5)^{-2} =$

15) $\frac{120x^{18}y^6}{12x^8y^3} =$

16) $\frac{28x^{14}}{7x^8} =$

17) $\frac{33x^7y^4z^3}{11x^2y^4z} =$

18) $\frac{15x^2+10x}{5x} =$

19) $\frac{200x^5y^9}{100x^5y^8} =$

20) $(12x^2)(6x^2) =$

21) $\frac{36x^2y^4+18xy^6}{9xy} =$

22) $\frac{-48x^7y^{14}}{6x^5y^{11}} =$

23) $\frac{15x^6y^{15}}{10x^4y^4} =$

24) $\frac{36x^{16}y^{12}z}{9x^4y^5} =$

Binomial Operations

Solve each operation below.

1) $3x + 8 - (6x - 3) =$

2) $(4x - 5) + (6x - 7) =$

3) $(-5x - 5) + (8x + 2) =$

4) $(4x - 1.4) + (5x - 3.6) =$

5) $\frac{1}{8}x + 4 - \left(\frac{1}{3}x - 5\right) =$

6) $5x + 3 - (7x - 2) =$

7) $14x + 5 - (26x - 1) =$

8) $(x + 8)(x + 5) =$

9) $(x - 7)(x - 3) =$

10) $(x - 4)(3x + 4) =$

11) $(x - 8)(x + 8) =$

12) $(x - 6)(9x + 5) =$

13) $(4x - 5)(4x + 5) =$

14) $(x + 9)(x - 6) =$

15) $(x - 7)(5x + 7) =$

16) $(x^2 + 6)(x^2 - 6) =$

17) $(x - 3)(x + 3) =$

18) $7x(6x - 4) =$

19) $13x(3x + 5) =$

20) $(3x + 4) + (5x - 7) + (x - 9) =$

21) $(x^2 + 2)(x^2 - 2)$

22) $(3x - 3)(5x + 4)$

23) $(x - 5)(7x + 2)$

24) $(x - 3.4)(4.1x + 3.4)$

Polynomial Operations

Simplify each expression.

1) $(4x^2 + x - 6) + (2x - 4x^2 - 8) =$

2) $(3x^2 + 2x - 3) - (4x - 3x^2 - 2) =$

3) $(12x^2 - 8x + 3) - (-4x + 6x^2 - 3) =$

4) $(6x^5 - 2x^3 - 7x) + (5x + 14x^4 - 15) + (3x^2 + x^3 + 12) =$

5) $(15x^2 - 8x + 4) + (6x^2 - 2x + 3) =$

6) $12(3x^2 - 7x - 2) =$

7) $3x^3(3x^2 - 3x + 5) =$

8) $3x^2y^2(4x^2 - 6x + 3) =$

9) $(x + 5)(x^2 - 4x + 7) =$

10) $x(3x^2 - 3x + 7) =$

11) $5(x^2 - 3x + 6) =$

12) $(x - 3)(x^2 + 7x - 2) =$

13) $(12x^3 + 5x^2 - 13) + (-5x^3 + 3x^2 + 11) =$

14) $(3x - 4)(3x^2 + 5x + 10) =$

Squaring a Binomial

Write each square as a trinomial.

1) $(a + 2b)^2 =$

2) $(x + 5)^2 =$

3) $(2a - b)^2 =$

4) $(3x + 2)^2 =$

5) $(x - 8)^2 =$

6) $(2x + \frac{1}{4})^2 =$

7) $(4x - 5y)^2 =$

8) $(x - 7)^2 =$

9) $(x + 11)^2 =$

10) $(4x - 5)^2 =$

11) $(3x + 3y)^2 =$

12) $(3x + 8)^2 =$

13) $(3x + \frac{1}{3})^2 =$

14) $(2x^2 + 2y^2)^2 =$

15) $(x - 12)^2 =$

16) $(x + \sqrt{3})^2 =$

17) $(6x - 7)^2 =$

18) $2(x + 3)^2 =$

19) $(8x - 3y)^2 =$

20) $(x + 13)^2 =$

21) $2(x + 2)^2 =$

22) $(x^2 - 9)^2 =$

23) $(2x + 5)^2 =$

24) $(4x + 7)^2 =$

THEA Math Practice Book

Factor polynomial

Factor each completely.

1) $x^2 + 17x + 66 =$

2) $12x^2 - 36x =$

3) $x^3 - 7x^2 - 7x + 49 =$

4) $x^2 + 12x + 32 =$

5) $x^4 - 4x^2 - 12 =$

6) $x^2 - 14x + 45 =$

7) $10 + 6x + 32 + x =$

8) $3x^2 - 18x + 16x - 4 =$

9) $24x^3y + 8x^2y - 32xy =$

10) $20x^2 - 7x - 3 =$

11) $3x^2 - 26x + 35 =$

12) $\dfrac{3x^2 - 15x + 18}{x^2 - 9x + 14} =$

13) $\dfrac{(x-3)(x-5)}{(x-3)(x-9)} =$

14) $(x - 4)4x + (x - 4)4 =$

15) $8x^2 - 20x^4 =$

16) $\dfrac{x^2 + 10x + 24}{(x+4)} =$

17) $x^2 + 2x - 63 =$

18) $4x^4 + 24x^2 - 28x^3 - 168x =$

19) $12(a - b) - 4a(a - b) =$

20) $18x^2 - 18x =$

Answer key Chapter 7

Adding and Subtracting Monomial

1) $13x^3$
2) $7x^2$
3) $\frac{5}{8}x^3$
4) $9x^4$
5) $6x^7$
6) $4x^3$
7) 0
8) $2x^{20}$
9) $5x^{-5}$
10) $20p^8$
11) $-1.8x^2$
12) $7\frac{3}{5}x^4$
13) $2x^{13}$
14) $4p^6$
15) $-5p^4$
16) $6.2x^6$
17) $\frac{7}{5}x^8$
18) $11x^4$
19) $-10.1x^2$
20) $-9x^7$
21) $12x^5$
22) $4x^{13}$
23) $-11x^{-10}$
24) $10x^{-7}$

Multiplying and Dividing Monomial

1) $10x^5y^3$
2) $28x^4y$
3) $-30x^4y^7$
4) $5x^9y^{12}$
5) $-60x^6$
6) $-15x^5y^7z^5$
7) $77x^{14}y^{16}$
8) $-36x^4y^9$
9) $-48x^3$
10) $-24x^5y^9$
11) $-22x^{-11}y^3$
12) $30x^{11}z^2$
13) $2x^6y^2$
14) $\frac{1}{36}y^{-10}$
15) $10x^{10}y^3$
16) $4x^6$
17) $3x^5z^4$
18) $3x + 2$
19) $2y$
20) $72x^4$
21) $4xy^3 + 2y^5$
22) $-8x^2y^3$
23) $\frac{3}{2}x^2y^{11}$
24) $4x^{12}y^7z$

Binomial Operations

1) $-3x + 11$
2) $10x - 12$
3) $3x - 3$
4) $9x - 5$
5) $-\frac{5}{24}x + 9$
6) $-2x + 5$
7) $-12x + 6$
8) $x^2 + 13x + 40$
9) $x^2 - 10x + 21$
10) $3x^2 - 8x - 16$
11) $x^2 - 64$
12) $9x^2 - 49x - 30$
13) $16x^2 - 25$
14) $x^2 + 3x - 54$
15) $5x^2 - 28x - 49$
16) $x^4 - 36$
17) $x^2 - 9$
18) $42x^2 - 28x$
19) $39x^2 + 65x$
20) $9x - 12$
21) $x^4 - 4$
22) $15x^2 - 3x - 12$
23) $7x^2 - 33x - 10$
24) $4.1x^2 - 10.54x - 11.56$

THEA Math Practice Book

Polynomial Operations

1) $3x - 14$
2) $6x^2 - 2x - 1$
3) $6x^2 - 4x + 6$
4) $6x^5 + 14x^4 - x^3 + 3x^2 - 3$
5) $21x^2 - 10x + 7$
6) $36x^2 - 84x - 24$
7) $9x^5 - 9x^4 + 15x^3$
8) $12x^4y^2 - 18x^3y^2 + 9x^2y^2$
9) $x^3 + x^2 - 13x + 35$
10) $3x^3 - 3x^2 + 7x$
11) $5x^2 - 15x + 30$
12) $x^3 + 4x^2 - 23x + 6$
13) $7x^3 + 8x^2 - 2$
14) $9x^3 + 3x^2 + 10x - 40$

Squaring a Binomial

1) $a^2 + 4b^2 + 4ab$
2) $x^2 + 10x + 25$
3) $4a^2 + b^2 - 4ab$
4) $9x^2 + 12x + 4$
5) $x^2 - 16x + 64$
6) $x^2 + x + \frac{1}{16}$
7) $16x^2 - 40xy + 25y^2$
8) $x^2 - 14x + 49$
9) $x^2 + 22x + 121$
10) $16x^2 - 40x + 25$
11) $9x^2 + 18xy + 9y^2$
12) $9x^2 + 48x + 64$
13) $9x^2 + 2x + \frac{1}{9}$
14) $4x^4 + 4y^4 + 8x^2y^2$
15) $x^2 - 24x + 144$
16) $x^2 + 2\sqrt{3}x + 3$
17) $36x^2 - 84x + 49$
18) $2x^2 + 12x + 18$
19) $64x^2 - 48xy + 9y^2$
20) $x^2 + 36x + 169$
21) $2x^2 + 8x + 8$
22) $x^4 - 18x^2 + 81$
23) $4x^2 + 20x + 25$
24) $16x^2 + 56x + 49$

Factor polynomial

1) $(x + 6)(x + 11)$
2) $12x(x - 3)$
3) $(x^2 - 7)(x - 7)$
4) $(x + 8)(x + 4)$
5) $(x^2 - 6)(x^2 + 2)$
6) $(x - 9)(x - 5)$
7) $7(x + 6)$
8) $3x(x - 2) + 4(x - 1)$
9) $8xy(3x^2 + x - 4)$
10) $(4x + 1)(5x - 3)$
11) $(3x - 5)(x - 7)$
12) $\frac{3(x-3)}{x-7}$
13) $\frac{x-5}{x-9}$
14) $(x - 4)(4x + 4)$
15) $-2x^2(-4 + 10x^2)$
16) $x + 6$
17) $(x + 9)(x - 7)$
18) $4x(x^2 + 6)(x - 7)$
19) $(a - b)(12 - 4a)$
20) $18x(x - 1)$

WWW.MathNotion.com

Chapter 8:
Functions

Relation and Functions

Determine whether each relation is a function. Then state the domain and range of each relation.

1)
Function: ....
Domain: ....
Range: ....

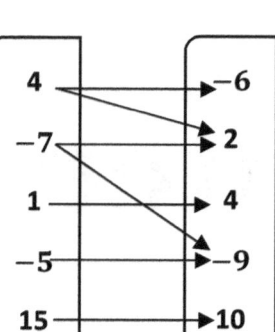

2)
Function: ....
Domain: ....
Range: ....

x	y
4	5
2	3
−6	−8
6	−8
−11	2

3)
Function: ....
Domain: ....
Range: ....

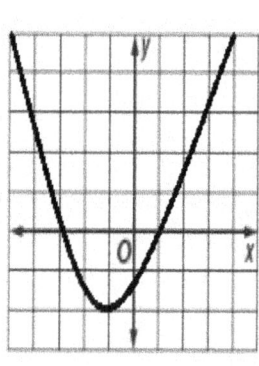

4) $\{(2,-2),(7,-6),(9,9),(8,1),(7,4)\}$
Function: ....
Domain: ....
Range: ....

5)
Function: ....
Domain: ....
Range: ....

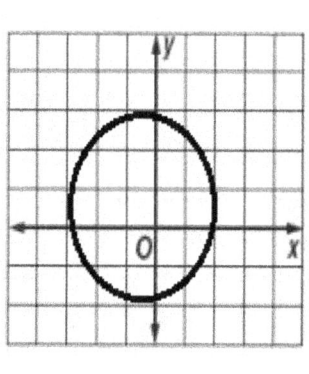

6)
Function: ....
Domain: ....
Range: ....

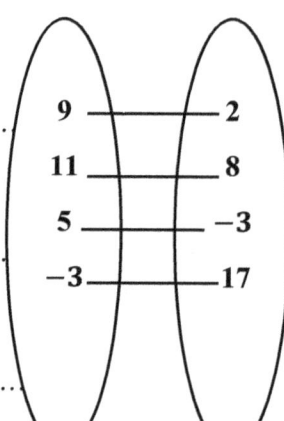

THEA Math Practice Book

Slope form

Write the slope-intercept form of the equation of each line.

1) $4x + 5y = 15$

2) $4x + 12y = 3$

3) $7x + y = -9$

4) $-7x + 11y = 5$

5) $5x - 4y = 7$

6) $-21x + 3y = 6$

7) $2x + y = 0$

8) $5x - 7y = -9$

9) $-13.5x + 27y = 54$

10) $-3x + \frac{2}{3}y = 18$

11) $10x + y = -120$

12) $3x = -36y - 27$

13) $1.5x = 3y + 3$

14) $5x = -\frac{5}{4}y + 25$

Slope and Y-Intercept

Find the slope and y-intercept of each equation.

1) $y = \frac{1}{5}x + 4$

2) $y = 7x + 8$

3) $x - 3y = 9$

4) $y = 5x + 21$

5) $y = 9$

6) $y = -2x + 3$

7) $x = -15$

8) $y = 7x$

9) $y - 3 = 4(x + 1)$

10) $x = -\frac{5}{8}y - \frac{1}{3}$

WWW.MathNotion.com

Slope and One Point

Find a Point-Slope equation for a line containing the given point and having the given slope.

1) $m = -2, (1, -1)$

2) $m = 3, (1, 2)$

3) $m = -2, (-1, -5)$

4) $m = 1, (3, 2)$

5) $m = 5, (2, 4)$

6) $m = \frac{3}{2}, (4, 5)$

7) $m = 0, (-4, -5)$

8) $m = 2, (1, -3)$

9) $m = 1, (0, 3)$

10) $m = \frac{3}{4}, (-2, -5)$

11) $m = -3, (1, -1)$

12) $m = -2, (2, -1)$

13) $m = 5, (1, 0)$

14) $m =$ undefined, $(8, -8)$

15) $m = -\frac{1}{8}, (8, 4)$

16) $m = \frac{1}{4}, (3, 2)$

17) $m = -8, (2, 4)$

18) $m = 6, (-2, -4)$

19) $m = \frac{1}{3}, (3, 1)$

20) $m = \frac{-4}{9}, (0, -3)$

21) $m = \frac{1}{4}, (4, 4)$

22) $m = -5, (0, -1)$

23) $m = 0, (0.9, -3)$

24) $m = -\frac{5}{7}, (7, -1)$

25) $m = 0, (-4, 8)$

26) $m =$ Undefined, $(-10, -2)$

Slope of Two Points

Write the slope-intercept form of the equation of the line through the given points.

1) $(1, 0), (-1, 5)$

2) $(-1, 3), (5, 6)$

3) $(-5, 1), (-1, 5)$

4) $(2, -3), (-9, 8)$

5) $(5, 0), (3, 1)$

6) $(9, -1), (-1, 9)$

7) $(-5, 3), (-6, 1)$

8) $(-7, -2), (1, 0)$

9) $(-5, -5), (3, 3)$

10) $(-1, 9), (-1, -5)$

11) $(-2, 7), (1, 7)$

12) $(1, -5), (4, -4)$

13) $(6, -9), (-3, 0)$

14) $(1, -4), (7, 4)$

15) $(-9, 5), (-3, -1)$

16) $(9, 5), (5, 1)$

17) $(10, -7), (2, -6)$

18) $(-5, -9), (-7, 2)$

19) $(7, 4), (3, 1)$

20) $(-1, -1), (9, 2)$

21) $(-8, 8), (8, 2)$

22) $(9, 2), (5, 11)$

23) $(8, 2), (9, 3)$

24) $(-2, -5), (-5, -8)$

Equation of Parallel and Perpendicular lines

Write the slope-intercept form of the equation of the line described.

1) Through: $(-5, 2)$, parallel to $y = 2x + 5$

2) Through: $(-4, 1)$, parallel to $y = -3x$

3) Through: $(-10, -2)$, perpendecular to $y = \frac{1}{2}x + 8$

4) Through: $(6, -2)$, parallel to $y = -5x + 13$

5) Through: $(-7, 4)$, parallel to $y = \frac{3}{7}x - 6$

6) Through: $(2, 0)$, perpendecular to $y = -\frac{1}{5}x + 8$

7) Through: $(4, -7)$, perpendecular to $y = -6x - 10$

8) Through: $(-5, 1)$, perpendecular to $y = -\frac{1}{8}x + 3$

9) Through: $(-1, -2)$, parallel to $2y + 4x = 9$

10) Through: $(1, 10)$, parallel to $y = \frac{1}{10}x - 5$

11) Through: $(5, -5)$, parallel to $y = 9$

12) Through: $(7, 2)$, perpendecular to $y = \frac{5}{2}x + 3$

13) Through: $(0, -4)$, perpendecular to $3y - x = 11$

14) Through: $(3, 5)$, parallel to $3y + x = 5\frac{3}{4}$

15) Through: $(1, 1)$, perpendecular to $y = 5x + 12$

16) Through: $(-3, -5)$, parallel to $8y - x = 10$

17) Through: $(-2, -2)$, perpendecular to $y = 4x + \frac{1}{7}$

18) Through: $(-8, 0)$, perpendecular to $5y - 4x - 9 = 0$

Quadratic Equations - Square Roots Law

Solve each equation by taking square roots.

1) $x^3 - 4 = 4$

2) $x^2 - 6 = 19$

3) $12x^2 - 2 = 100$

4) $-x^2 - 2 = -66$

5) $6x^2 + 3 = 489$

6) $5x^2 + 9 = 14$

7) $8x^2 - 17 = 2{,}791$

8) $7x^2 + 16 = 2{,}151$

9) $100x^2 = 4$

10) $4x^2 - 8 = 68$

11) $9x^2 - 5 = 607$

12) $20x^2 - 20 = 60$

13) $13x^2 - 3 = 4{,}209$

14) $13x^2 - 8 = -1{,}139$

15) $16x^2 - 16 = 48$

16) $54x^2 = 6$

17) $-8x^2 - 8 = -31$

18) $15x^2 - 30 = 255$

19) $42x^2 = -126$

20) $16x^2 - 10 = 39$

21) $2x^2 + 16 = 160$

22) $13x^2 = 117$

23) $-16x^2 + 12 = 340$

24) $6x^2 - 30 = -24$

25) $12x^2 - 60 = -48$

26) $3x^2 + 15 = 315$

27) $27x^2 + 3 = 975$

28) $21x^2 + 3 = 87$

Quadratic Equations - Factoring

Solve each equation by factoring.

1) $(10n - 5)(9n + 3) = 0$

2) $(20n - 4)(4n + 4) = 0$

3) $x^2 + 16 = 10x$

4) $7x^2 - 42 = -35x$

5) $2x^2 - 7 = 13x$

6) $7x^2 + 32 = 7 - 40x$

7) $-6x^2 + 2x + 144 = 6x^2 + 14x$

8) $8x^2 + 3x + 2 = 7x^2$

9) $8x^2 - 57x = -7$

10) $21x^2 + 75 = -120x$

11) $3n(2n - 10) = 0$

12) $(n + 4)(5n - 6) = 0$

13) $x^2 + 9 = -9 - 9x$

14) $5x^2 + 12 = -20x - 8$

15) $3x^2 + 20 = -18x - 7$

16) $5x^2 + 2x = 10 - 3x$

17) $3x^2 - 8x + 4 = 20$

18) $2x^2 + 40 = -24x - 32$

19) $3x^2 - 8x = 16$

20) $3x^2 + 50 = -30x - 25$

21) $6x^2 - 16x - 32 = 0$

22) $5x^2 + 5x - 50 = 10$

23) $2x^2 + 6x = -10x - 32$

24) $2n(n + 2) = 0$

Quadratic Equations - Completing the Square

Solve each equation by completing the square.

1) $3x^2 - 6x - 9 = 0$

2) $5x^2 - 10x - 15 = 0$

3) $2x^2 + \frac{5x}{2} - 3 = 0$

4) $-3x^2 + 2x + 8 = 0$

5) $3x^2 + 42x - 153 = 0$

6) $3x^2 + 18x + 4 = -20$

7) $x^2 + 14x - 2 = 13$

8) $x^2 - 4x - 91 = 7$

9) $2x^2 - 18x = -36$

10) $5x^2 = -20x + 60$

11) $12x^2 = 48x - 36$

12) $\frac{1}{4}x^2 - \frac{2}{4}x - \frac{3}{4} = 0$

13) $3x^2 - 36x + 25 = -8$

14) $3x^2 - 30x + 54 = 0$

15) $x^2 + 6x = 59$

16) $3x^2 = 54x + 120$

17) $\frac{1}{5}x^2 + \frac{2}{5}x = -4$

18) $2x^2 - 24x = 22$

19) $6x^2 - 12x - 18 = 0$

20) $3x^2 + 42x - 45 = 0$

21) $\frac{1}{2}x^2 - x - \frac{3}{2} = 0$

22) $7x^2 - 28x = -21$

23) $9x^2 + 54x + 72 = 0$

24) $-7x^2 - 42x = 56$

Quadratic Equations - Quadratic Formula

Solve each equation with the quadratic formula.

1) $3x^2 + 6x - 24 = 0$

2) $4x^2 - 8x = -4$

3) $\frac{1}{4}x^2 = \frac{9}{4}x - 5$

4) $\frac{2}{3}x^2 + \frac{10}{3}x - 4 = 0$

5) $3x^2 = 27x - 60$

6) $3x^2 - 12x - 26 = 10$

7) $7x^2 = -21x + 280$

8) $3x^2 + 15x - 18 = 0$

9) $x^2 + x - 2 = \frac{1}{4}$

10) $\frac{4}{3}x^2 - \frac{2}{3}x = 3$

11) $x^2 = -3x + 40$

12) $9x^2 - 25 = 8x$

13) $\frac{8}{5}x^2 - \frac{8}{5}x = 3$

14) $6x^2 - 3x - 10 = 8$

15) $\frac{1}{3}x^2 = 3x - \frac{100}{15}$

16) $10x^2 + 30x = 400$

17) $x^2 - 2 = \frac{1}{8}x$

18) $24x^2 + 18x + 15 = 0$

19) $x^2 - \frac{1}{2}x - \frac{13}{2} = 1$

20) $11x^2 + 1 = 5x^2 + 7x$

21) $17x^2 + 14 = x$

22) $10x^2 - 5x - 20 = 10$

Arithmetic Sequences

Find the three terms in the sequence after the last one given.

1) $2, 6, 10, 14, ..., ..., ...,$

2) $-10, -6, -2, 2, ..., ..., ...,$

3) $a_1 = -12, d = 12$

4) $a_1 = -10.6, d = 1.4$

5) $a_{10} = 36, d = 6$

6) $a_n = (3n)^2$

7) $24, 16, 8, 0, ..., ..., ...,$

8) $-15, -30, -45, -60, ..., ..., ...,$

9) $a_n = -18 + 3n$, find a_{26}

10) $a_n = -8.4 - 5.8n$, find a_{32}

11) $a_5 = \frac{3}{8}, d = -\frac{1}{4}$

12) $a_n = \frac{2n^2}{3n+2}$

13) $-8, -15, -22, ..., ..., ...,$

14) $a_n = 4n + 3$

15) $-8, -3, 2, 7, ..., ..., ...,$

16) $-12, -7, -2, 3, ..., ..., ...,$

17) $\frac{6}{7}, \frac{5}{14}, -\frac{1}{7}, -\frac{9}{14}, ..., ..., ...,$

18) $5, -5, -15, ..., ..., ...,$

19) $9, 13, 17, ..., ..., ...,$

20) $11, 22, 33, ..., ..., ...,$

21) $-3.5, -4.7, -5.9, ..., ..., ...,$

22) $115, 140, 165, 190, ..., ..., ...,$

23) $-\frac{8}{9}, -\frac{5}{9}, -\frac{2}{9}, ..., ..., ...,$

24) $-2.25, -1.50, -0.75, ..., ..., ...,$

THEA Math Practice Book

Geometric Sequences

Find the three terms in the sequence after the last one given.

1) $3, 9, 27, 81, \ldots, \ldots, \ldots$

2) $320, 80, 20, \ldots, \ldots, \ldots$

3) $a_n = a_{n-1} \times (-2), a_1 = 1$

4) $a_n = a_{n-1} \times 4, a_1 = 3$

5) $0.2, 0.6, 1.8, 5.4 \ldots, \ldots, \ldots$

6) $a_n = 2^{n-1}, a_1 = 1$

7) $a_n = 8(\frac{1}{4})^{n-1}, a_1 = 8$

8) $1, \frac{3}{2}, \frac{9}{4}, \frac{27}{8}, \ldots \ldots, \ldots \ldots, \ldots$

9) $a_n = 2a_{n-1}, a_1 = 5$

10) $1, 2, 4, \ldots, \ldots, \ldots$

11) $-4, -12, -36, \ldots, \ldots, \ldots$

12) $-0.2, 1, -5, \ldots, \ldots, \ldots$

13) $a_n = (-2)^{2n+1}$

14) $a_n = 0.25 \times (4)^{n-1}$ find a_5

15) $a_n = 3 \times (2)^{n-1}$; $a_6=?$

16) $3, 1, \frac{1}{3}, \ldots, \ldots, \ldots, \ldots$

17) $2, 6, 18, \ldots, \ldots, \ldots$

18) $5, -5, 5, \ldots, \ldots, \ldots$

19) $243, 162, 108, \ldots, \ldots, \ldots$

20) $-1.5, 3, -6, \ldots, \ldots, \ldots$

21) $-0.25, -1, -4, \ldots, \ldots, \ldots$

22) $1, -6, 36, \ldots, \ldots, \ldots$

23) $a_n = -0.1(-2)^{n-1}$

24) $a_n = -0.5 \times 4^{n-1}$

www.MathNotion.com

Answer key Chapter 8

Relation and Functions

1) No, $D_f = \{4, -7, 1, -5, 15\}$, $R_f = \{-6, 2, 4, -9, 10\}$
2) Yes, $D_f = \{4, 2, -6, 6, -11\}$, $R_f = \{5, 3, -8, 2\}$
3) Yes, $D_f = (-\infty, \infty)$, $R_f = \{-2, \infty)$
4) No, $D_f = \{2, 7, 9, 8, 7\}$, $R_f = \{-2, -6, 9, 1, 4\}$
5) No, $D_f = [-3, 2]$, $R_f = [-2, 3]$
6) Yes, $D_f = \{9, 11, 5, -3\}$, $R_f = \{2, 8, -3, 17\}$

Slope form

1) $y = -\frac{4}{5}x + 3$
2) $y = -\frac{1}{3}x + \frac{1}{4}$
3) $y = -7x - 9$
4) $y = \frac{7}{11}x + \frac{5}{11}$
5) $y = \frac{5}{4}x - \frac{7}{4}$
6) $y = 7x + 2$
7) $y = -2x$
8) $y = \frac{5}{7}x + \frac{9}{7}$
9) $y = 0.5x + 2$
10) $y = 4.5x + 27$
11) $y = -10x - 120$
12) $y = -\frac{1}{12}x - \frac{3}{4}$
13) $y = 0.5x - 1$
14)

Slope and Y-Intercept

1) $m = \frac{1}{5}, b = 4$
2) $m = 7, b = 8$
3) $m = \frac{1}{3}, b = -3$
4) $m = 5, b = 21$
5) $m = 0, b = 9$
6) $m = -2, b = 3$
7) $m = undefind$, $b: no\ intercept$
8) $m = 7, b = 0$
9) $m = 4, b = 7$
10) $m = -\frac{8}{5}, b = -\frac{1}{3}$

Slope and One Point

1) $y = -2x + 1$
2) $y = 3x - 1$
3) $y = -2x - 7$
4) $y = x - 1$
5) $y = 5x - 6$
6) $y = \frac{3}{2}x - 1$
7) $y = -5$
8) $y = 2x - 5$
9) $y = x + 3$
10) $y = \frac{3}{4}x - \frac{7}{2}$
11) $y = -3x + 2$
12) $y = -2x + 3$
13) $y = 5x$
14) $x = 8$
15) $y = -\frac{1}{8}x + 5$
16) $y = \frac{1}{4}x + \frac{5}{4}$
17) $y = -8x + 20$
18) $y = 6x + 8$
19) $y = \frac{1}{3}x$
20) $y = -\frac{4}{9}x - 3$

21) $y = \frac{1}{4}x + 3$
22) $y = -5x - 1$

23) $y = -3$
24) $y = -\frac{5}{7}x + 4$

25) $y = 8$
26) $x = -10$

Slope of Two Points

1) $y = -\frac{5}{2}x + \frac{5}{2}$
2) $y = \frac{1}{2}x + \frac{7}{2}$
3) $y = x + 6$
4) $y = -x - 1$
5) $y = -\frac{1}{2}x + \frac{5}{2}$
6) $y = -x + 8$
7) $y = 2x + 13$
8) $y = \frac{1}{4}x - \frac{1}{4}$

9) $y = x$
10) $x = -1$
11) $y = 7$
12) $y = \frac{1}{3}x - 5\frac{1}{3}$
13) $y = -x - 3$
14) $y = \frac{4}{3}x - 5\frac{1}{3}$
15) $y = -x - 4$
16) $y = x - 4$

17) $y = -\frac{1}{8}x - 5\frac{3}{4}$
18) $y = -5\frac{1}{2}x - 36\frac{1}{2}$
19) $y = \frac{3}{4}x - 1\frac{1}{4}$
20) $y = \frac{3}{10}x - \frac{7}{10}$
21) $y = -\frac{3}{8}x + 5$
22) $y = -\frac{9}{4}x + 22\frac{1}{4}$
23) $y = x - 6$
24) $y = x - 3$

Equation of Parallel and Perpendicular lines

1) $y = 2x + 12$
2) $y = -3x - 11$
3) $y = -2x - 22$
4) $y = -5x + 28$
5) $y = \frac{3}{7}x + 7$
6) $y = 5x - 10$

7) $y = \frac{1}{6}x - 7\frac{2}{3}$
8) $y = 8x + 41$
9) $y = -2x - 4$
10) $y = \frac{1}{10}x + 9\frac{9}{10}$
11) $y = -5$
12) $y = -\frac{2}{5}x + 4\frac{4}{5}$

13) $y = -3x - 4$
14) $y = -\frac{1}{3}x + 6$
15) $y = -\frac{1}{5}x + 1\frac{1}{5}$
16) $y = \frac{1}{8}x - 4\frac{5}{8}$
17) $y = -\frac{1}{4}x - 2\frac{1}{2}$
18) $y = -\frac{5}{4}x - 10$

Quadratic Equations - Square Roots Law

1) 2
2) $\{5, -5\}$
3) $\{\frac{\sqrt{17}}{2}, -\frac{\sqrt{17}}{2}\}$
4) $\{8, -8\}$
5) $\{9, -9\}$
6) $\{1, -1\}$
7) $\{3\sqrt{39}, -3\sqrt{39}\}$

8) $\{\sqrt{305}, -\sqrt{305}\}$
9) $\{\frac{1}{5}, -\frac{1}{5}\}$
10) $\{\sqrt{19}, -\sqrt{19}\}$
11) $\{2\sqrt{17}, -2\sqrt{17}\}$
12) $\{2, -2\}$
13) $\{18, -18\}$
14) $\{i\sqrt{87}, -i\sqrt{87}\}$

15) $\{2, -2\}$
16) $\{\frac{1}{3}, -\frac{1}{3}\}$
17) $\{\frac{\sqrt{46}}{4}, -\frac{\sqrt{46}}{4}\}$
18) $\{\sqrt{19}, -\sqrt{19}\}$
19) $\{i\sqrt{3}, -i\sqrt{3}\}$
20) $\{\frac{7}{4}, -\frac{7}{4}\}$
21) $\{6\sqrt{2}, -6\sqrt{2}\}$

22) $\{3, -3\}$
23) $\{i\sqrt{\frac{41}{2}}, -i\sqrt{\frac{41}{2}}\}$
24) $\{1, -1\}$
25) $\{1, -1\}$
26) $\{10, -10\}$
27) $\{6, -6\}$
28) $\{2, -2\}$

Quadratic Equations - Factoring

1) $\{\frac{1}{2}, -\frac{1}{3}\}$
2) $\{\frac{1}{5}, -1\}$
3) $\{2, 8\}$
4) $\{-6, 1\}$
5) $\{-\frac{1}{2}, 7\}$
6) $\{-\frac{5}{7}, -5\}$
7) $\{3, -4\}$
8) $\{-2, -1\}$
9) $\{\frac{1}{8}, 7\}$
10) $\{-\frac{5}{7}, -5\}$
11) $\{5, 0\}$
12) $\{-4, \frac{6}{5}\}$
13) $\{-6, -3\}$
14) $\{-2\}$
15) $\{-3\}$
16) $\{-2, 1\}$
17) $\{-\frac{4}{3}, 4\}$
18) $\{-6\}$
19) $\{-\frac{4}{3}, 4\}$
20) $\{-5\}$
21) $\{-\frac{4}{3}, 4\}$
22) $\{-4, 3\}$
23) $\{-4\}$
24) $\{-2, 0\}$

Quadratic Equations - Completing the Square

25) $\{-1, 3\}$
26) $\{3, -1\}$
27) $\{-2, \frac{3}{4}\}$
28) $\{2, -\frac{4}{3}\}$
29) $\{-17, 3\}$
30) $\{-2, -4\}$
31) $\{-15, 1\}$
32) $\{2+\sqrt{102}, 2-\sqrt{102}\}$
33) $\{3, 6\}$
34) $\{-6, 2\}$
35) $\{3, 1\}$
36) $\{-1, 3\}$
37) $\{11, 1\}$
38) $\{5+\sqrt{7}, 5-\sqrt{7}\}$
39) $\{-3+2\sqrt{17}, -3-2\sqrt{17}\}$
40) $\{-2, 20\}$
41) $\{-1+i\sqrt{19}, -1-i\sqrt{19}\}$
42) $\{6+\sqrt{47}, 6-\sqrt{47}\}$
43) $\{3, -1\}$
44) $\{-15, 1\}$
45) $\{-1, 3\}$
46) $\{3, 1\}$
47) $\{-2, -4\}$
48) $\{-2, -4\}$

Quadratic Equations - Quadratic Formula

1) $\{2, -4\}$
2) $\{1\}$
3) $\{5, 4\}$
4) $\{1, -6\}$
5) $\{5, 4\}$
6) $\{6, -2\}$
7) $\{5, -8\}$
8) $\{1, -6\}$
9) $\{\frac{-1+\sqrt{10}}{2}, \frac{-1-\sqrt{10}}{2}\}$
10) $\{\frac{1+\sqrt{37}}{4}, \frac{1-\sqrt{37}}{4}\}$
11) $\{5, -8\}$
12) $\{\frac{4+\sqrt{241}}{9}, \frac{4-\sqrt{241}}{9}\}$
13) $\{\frac{2+\sqrt{34}}{4}, \frac{2-\sqrt{34}}{4}\}$
14) $\{2, -\frac{3}{2}\}$
15) $\{5, 4\}$
16) $\{5, -8\}$

17) $\{\frac{1+3\sqrt{57}}{16}, \frac{1-3\sqrt{57}}{16}\}$ 19) $\{3, -\frac{5}{2}\}$ 21) $\{\frac{1+i\sqrt{951}}{34}, \frac{1-i\sqrt{951}}{34}\}$

18) $\{\frac{-3+i\sqrt{31}}{8}, \frac{-3-i\sqrt{31}}{8}\}$ 20) $\{1, \frac{1}{6}\}$ 22) $\{2, -\frac{3}{2}\}$

Arithmetic sequences

1) $2, 6, 10, 14, 18, 22, 26$

2) $-10, -6, -2, 2, 6, 10, 14$

3) $-12, 0, 12, 24$

4) $-10.6, -9.2, -7.8, -6.4$

5) $-18, -12, -6, 0$

6) $9, 36, 81, 144$

7) $24, 16, 8, 0, -8, -16, -24$

8) $-15, -30, -45, -60, -75, -90, -105$

9) 60

10) -194

11) $\frac{11}{8}, \frac{9}{8}, \frac{7}{8}, \frac{5}{8}$

12) $\frac{2}{5}, 1, \frac{18}{11}, \frac{32}{14}$

13) $-8, -15, -22, -29, -36, -43$

14) $7, 11, 15, 19$

15) $-8, -3, 2, 7, 12, 17, 22$

16) $-12, -7, -2, 3, 8, 13, 18$

17) $\frac{6}{7}, \frac{5}{14}, -\frac{1}{7}, -\frac{9}{14}, -\frac{8}{7}, -\frac{23}{14}, -\frac{15}{7}$

18) $5, -5, -15, -25, -35, -45$

19) $9, 13, 17, 21, 25, 29$

20) $11, 22, 33, 44, 55, 66$

21) $-3.5, -4.7, -5.9 - 7.1, -8.3, -9.5$

22) $115, 140, 165, 190, 215, 240, 265$

23) $-\frac{8}{9}, -\frac{5}{9}, -\frac{2}{9}, \frac{1}{9}, \frac{4}{9}, \frac{7}{9}$

24) $-2.25, -1.50, -0.75, 0, 0.75, 1.5$

Geometric sequences

1) $3, 9, 27, 81, 243, 729, 2187$

2) $320, 80, 20, 5, 1.25, 0.3125$

3) $1, -2, 4, -8$

4) $3, 12, 48, 192$

5) $0.2, 0.6, 1.8, 5.4, 16.2, 48.6, 145.8$

6) $1, 2, 4, 8$

7) $8, 2, \frac{1}{2}, \frac{1}{8}$

8) $\frac{81}{16}, \frac{243}{32}, \frac{729}{64}$

9) $5, 10, 20, 40$

10) $1, 2, 4, 8, 16, 32$

11) $-4, -12, -36, -108, -324, -972$

12) $-0.2, 1, -5, 25, -125, 625$

13) $-8, -32, -128$

14) 64

15) 96

16) $3, 1, \frac{1}{3}, \frac{1}{9}, \frac{1}{27}, \frac{1}{81}$

17) $2, 6, 18, 54, 162, 486$

18) $5, -5, 5, -5, 5, -5$

19) $243, 162, 108, 72, 48, 32$

20) $-1.5, 3, -6, 12, -24, 48$

21) $-0.25, -1, -4, -16, -64, 256$

22) $1, -6, 36, -216, 1296, -7776$

23) $-0.1, 0.2, -0.4, 0.8$

24) $-0.5, -2, -8, -32$

Chapter 9:

Geometry

Area and Perimeter of Square

Find the perimeter and area of each squares.

1)

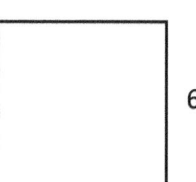

Perimeter:_____:

Area:_____:

2)

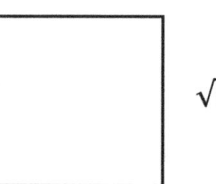

Perimeter:_____:

Area:_____:

3)

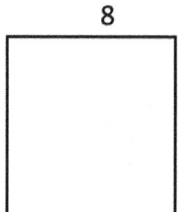

Perimeter:_____:

Area:_____:

4)

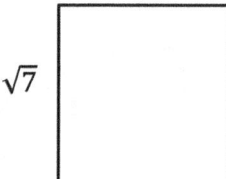

Perimeter:_____:

Area:_____:

5)

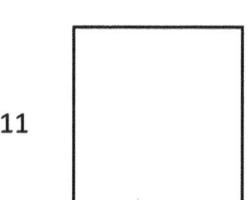

Perimeter:_____:

Area:_____:

6)

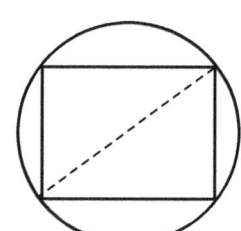

Perimeter of Square:_____:

Area of Square:_____:

Area and Perimeter of Rectangle

Find the perimeter and area of each rectangle.

1)

   7
2 ▭

Perimeter: _____.
Area: _____.

2)

   10
5 ▭

Perimeter: _____.
Area: _____.

3)

   13
7 ▭

Perimeter: _____.
Area: _____.

4)

   8
1.5 ▭

Perimeter: _____.
Area: _____.

5)

   3.6
2.4 ▭

Perimeter: _____.
Area: _____.

6)

   6
4 ▭

Perimeter: _____.
Area: _____.

Area and Perimeter of Triangle

Find the perimeter and area of each triangle.

1)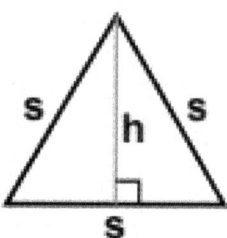

Perimeter: _____

Area: _____

2)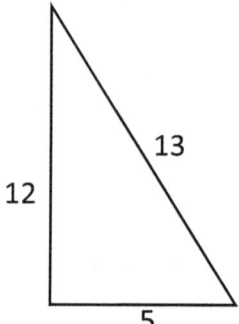

Perimeter: _____

Area: _____

3)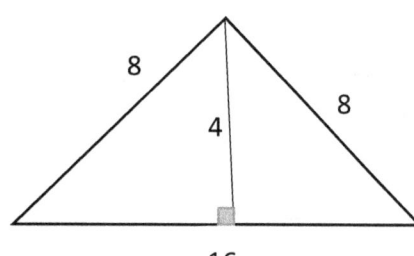

Perimeter: _____

Area _____

4) s=12

h=8

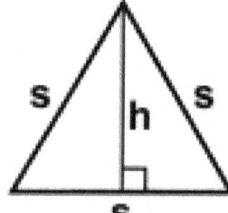

Perimeter: _____

Area: _____

5)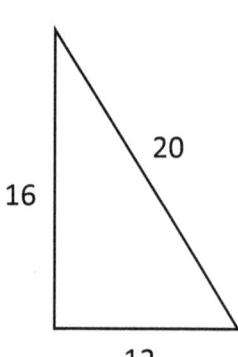

Perimeter: _____

Area: _____

6)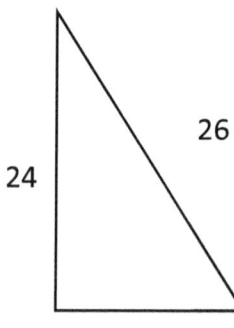

Perimeter: _____

Area: _____

Area and Perimeter of Trapezoid

Find the perimeter and area of each trapezoid.

1)
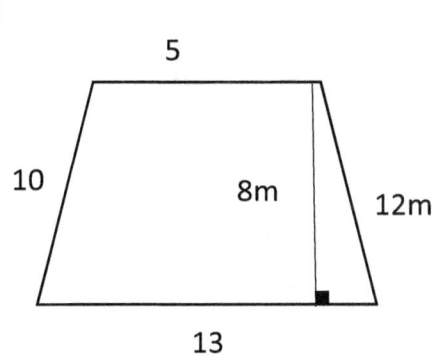

Perimeter: ....:

Area: ....:

2)
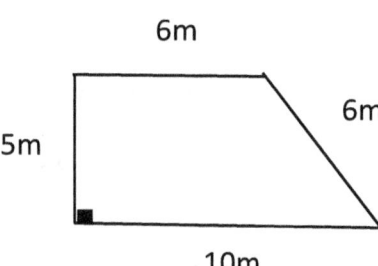

Perimeter: ....:

Area: ....:

3)

19

6   4

12

Perimeter: ....:

Area   ....:

4)
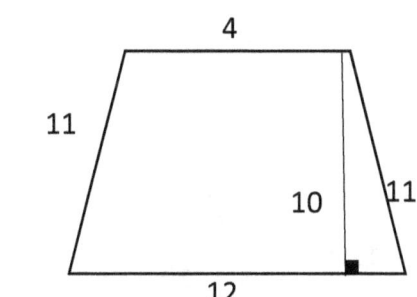

Perimeter: ....:

Area: ....:

5)

12

9   8   9

14

Perimeter: ....:

Area: ....:

6)
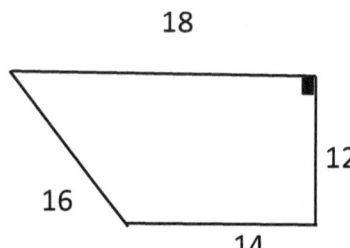

Perimeter: ....:

Area: ....:

Area and Perimeter of Parallelogram

Find the perimeter and area of each parallelogram.

1)

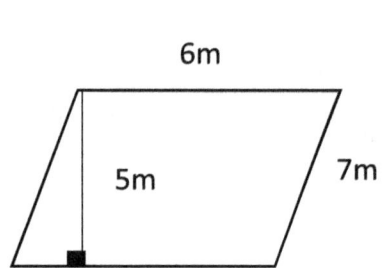

Perimeter:............:

Area:...............:

2)

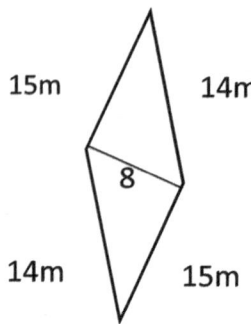

Perimeter:................:

Area:..............:

3)

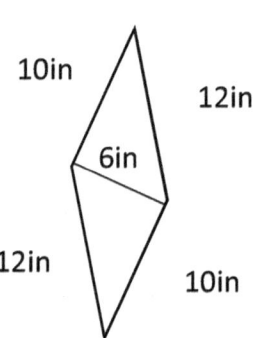

Perimeter:.............:

Area   ....:

4)

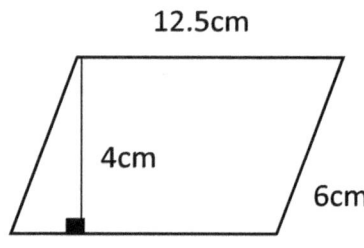

Perimeter:..............:

Area:..............:

5)

26m

14m

Perimeter:........:

Area:..........:

6)

15 m

Perimeter:.............:

Area:..............:

Circumference and Area of Circle

Find the circumference and area of each ($\pi = 3.14$).

1)

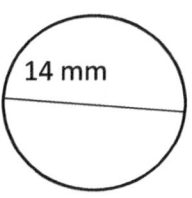

Circumference:

Area:

2)

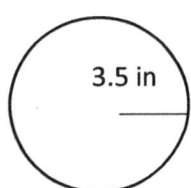

Circumference: _____.

Area: _____.

3)

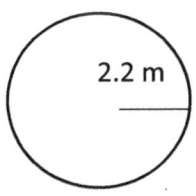

Circumference: _____.

Area _____.

4)

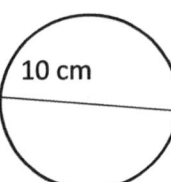

Circumference: _____.

Area: _____.

5)

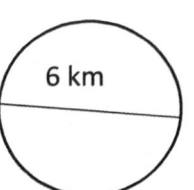

Circumference: _____.

Area: _____.

6)

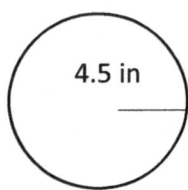

Circumference: _____.

Area: _____.

Perimeter of Polygon

Find the perimeter of each polygon.

1)

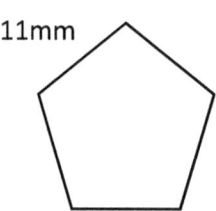

11mm

Perimeter: _____ :

2)

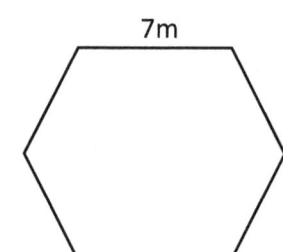

7m

Perimeter: _____ :

3)

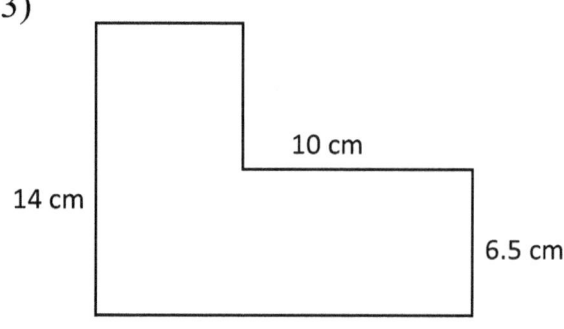

10 cm
14 cm
6.5 cm
18.5 cm

Perimeter: _____ :

4)

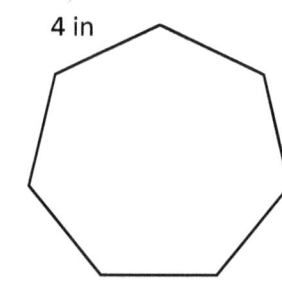

4 in

Perimeter: _____ :

5)

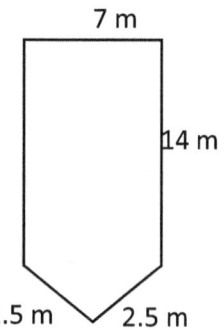

7 m
14 m
2.5 m   2.5 m

Perimeter: _____ :

6)

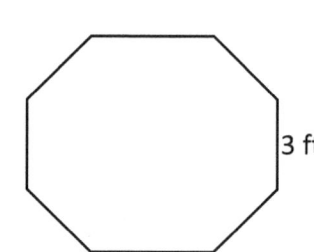

3 ft

Perimeter: _____ :

Volume of Cubes

Find the volume of each cube.

1)
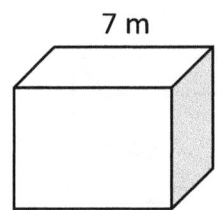

V:...................................

2)
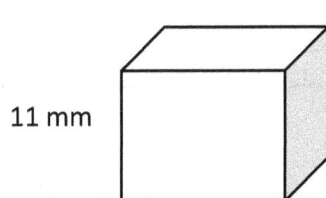

V:...................................

3)
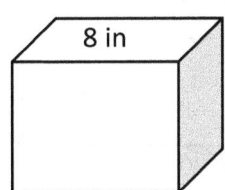

V:...................................

4)
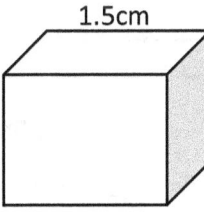

V:...................................

5)
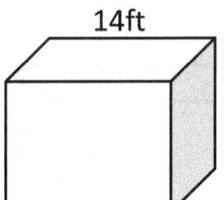

V:...................................

6)
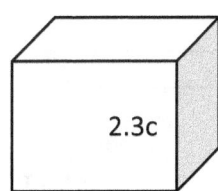

V:...................................

Volume of Rectangle Prism

Find the volume of each rectangle prism

1)

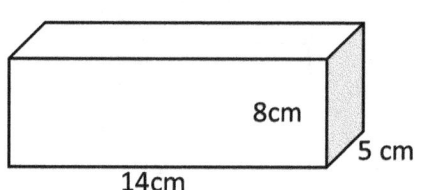

V:……………………………….

2)

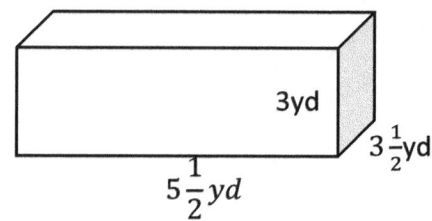

V:……………………………….

3)

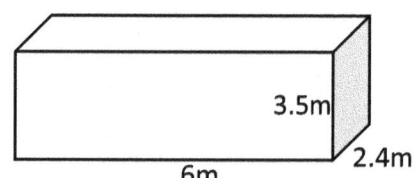

V:……………………………….

4)

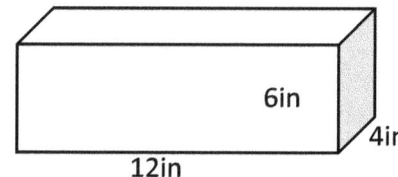

V:……………………………….

5)

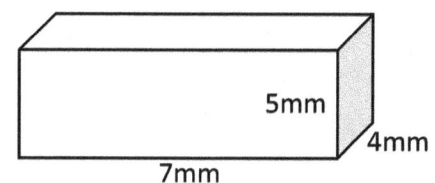

V:……………………………….

6)

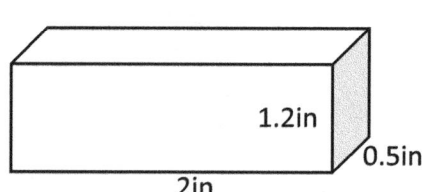

V:……………………………….

Volume of Cylinder

Find the volume of each cylinder.

1)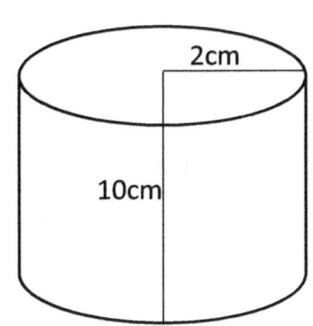

V: _____ .

2)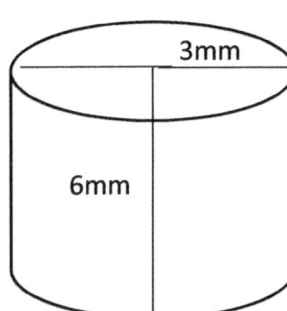

V: _____ .

3)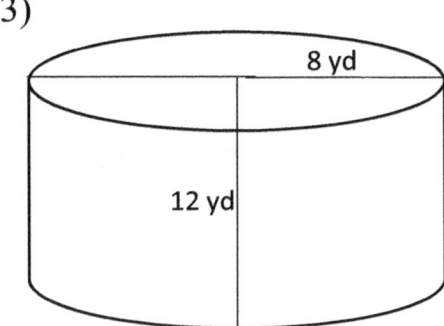

V: _____ .

4)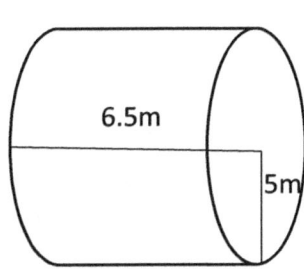

V: _____ .

5)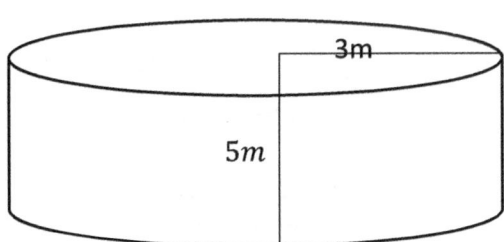

V: _____ .

6)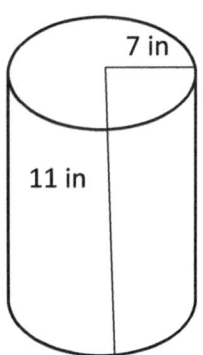

V: _____ .

Volume of Spheres

Find the volume of each spheres ($\pi = 3.14$).

1)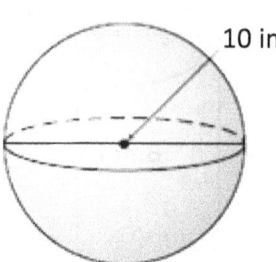

V:..............................

2)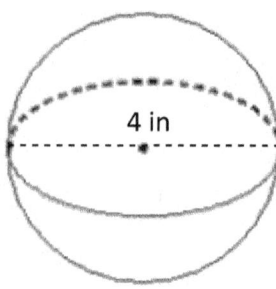

V:..............................

3)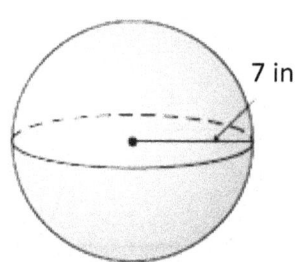

V:..............................

4)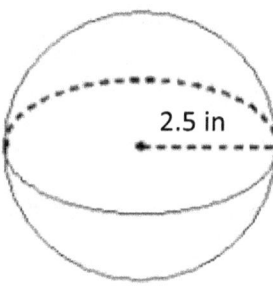

V:..............................

5)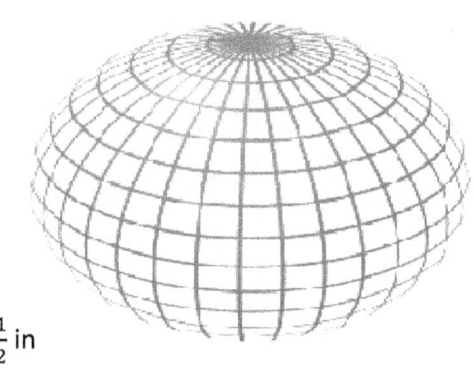

r = $4\frac{1}{2}$ in

V:..............................

6)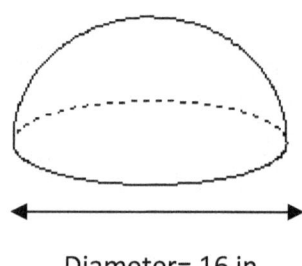

Diameter = 16 in

V:..............................

Volume of Pyramid and Cone

Find the volume of each pyramid and cone ($\pi = 3.14$).

1)

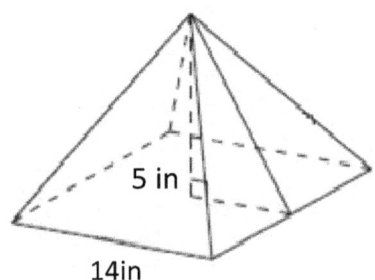

V:_____.

2)

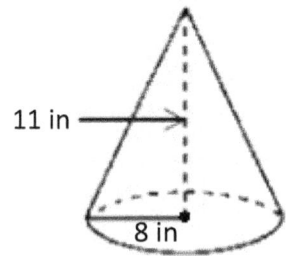

V:_____.

3)

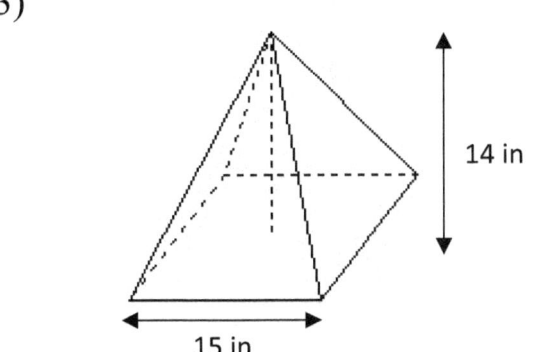

V:_____.

4)

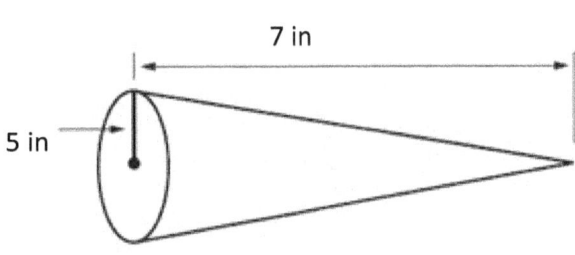

V:_____.

5)

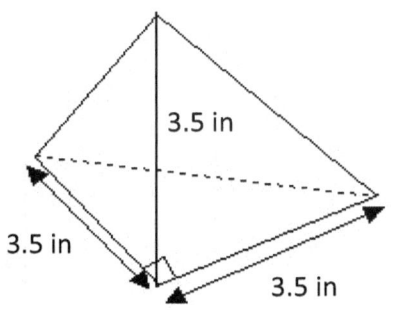

V:_____.

6)

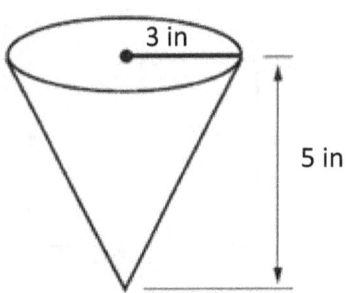

V:_____.

Surface Area Cubes

Find the surface area of each cube.

1)

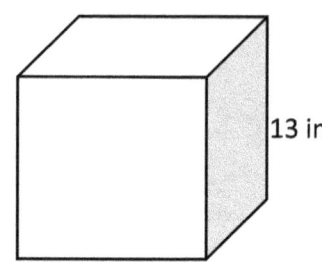

SA: _____.

2)

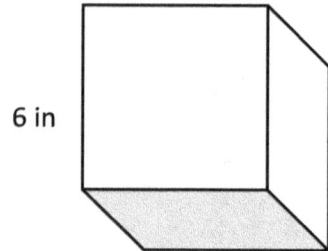

SA: _____.

3)

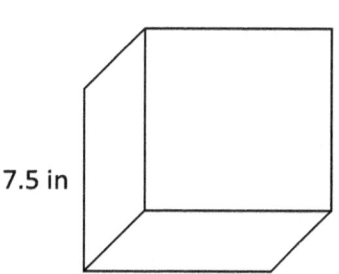

SA: _____.

4)

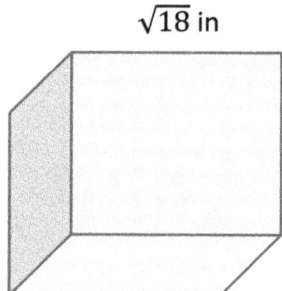

SA: _____.

5)

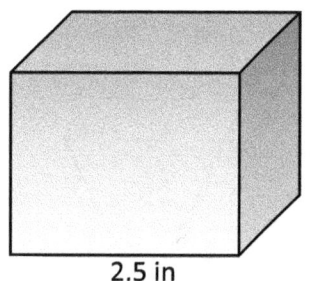

SA: _____.

6)

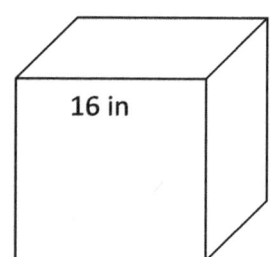

SA: _____.

Surface Area Rectangle Prism

Find the surface area of each rectangular prism.

1)
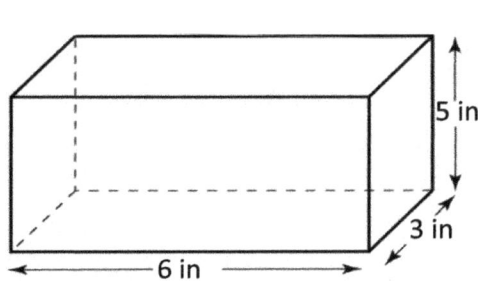
SA:_____.

2)
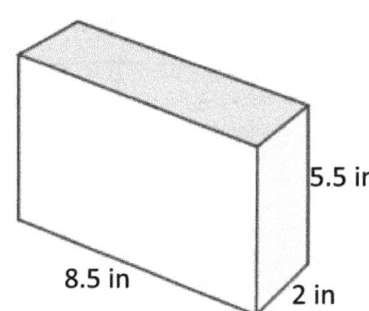
SA:_____.

3)
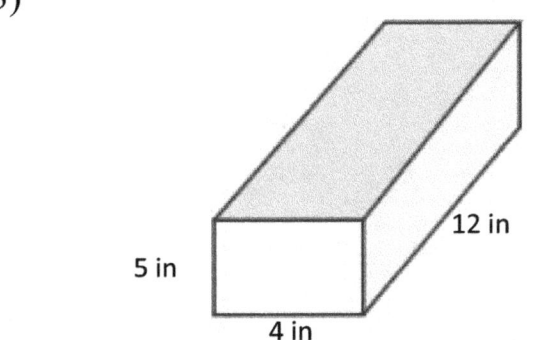
SA:_____.

4)
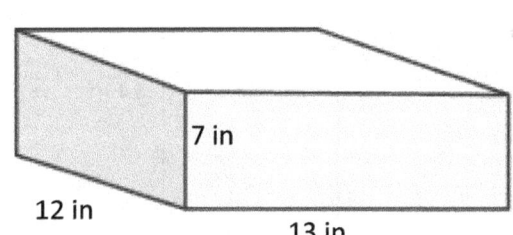
SA:_____.

5)
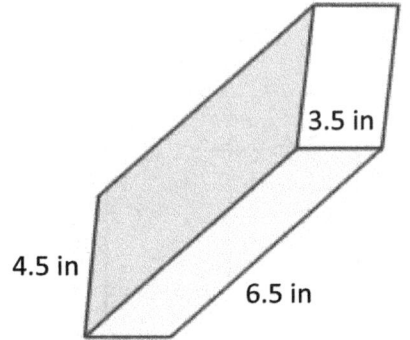
SA:_____.

6)
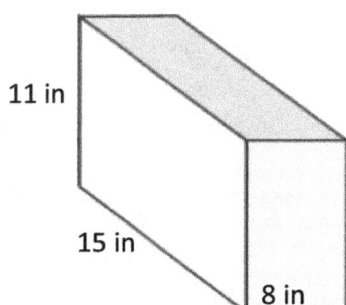
SA:_____.

Surface Area Cylinder

Find the surface area of each cylinder.

1)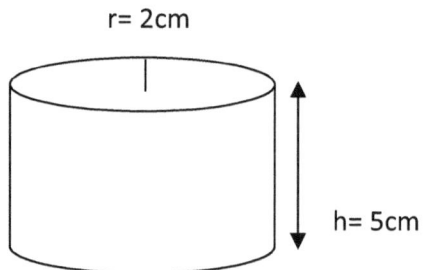

SA:_____.

2)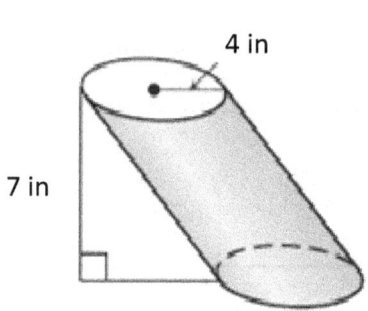

SA:_____.

3)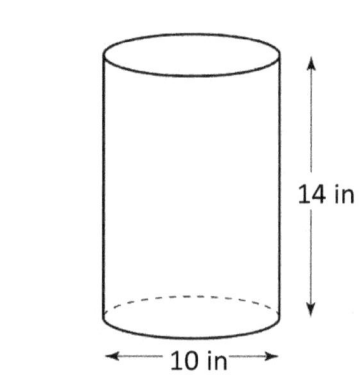

SA:_____.

4)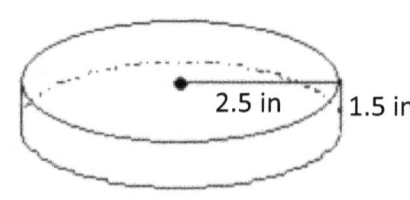

SA:_____.

5)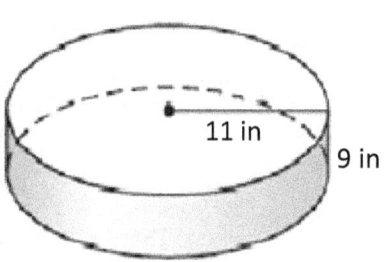

SA:_____.

6)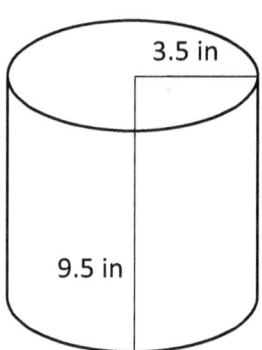

SA:_____.

Answer key Chapter 9

Area and Perimeter of Square

1. Perimeter: 24,   Area: 36
2. Perimeter: $4\sqrt{5}$, Area: 5
3. Perimeter: 32,   Area: 64
4. Perimeter: $4\sqrt{7}$, Area: 7
5. Perimeter: 44,   Area: 121
6. Perimeter: $4\sqrt{32}$, Area: 32

Area and Perimeter of Rectangle

1- Perimeter: 18,   Area: 14
2- Perimeter: 30,   Area: 50
3- Perimeter: 40,   Area: 91
4- Perimeter: 19, Area: 12
5- Perimeter: 12, Area: 8.64
6- Perimeter: 20,   Area: 24

Area and Perimeter of Triangle

1- Perimeter: 3s, Area: $\frac{1}{2}sh$
2- Perimeter: 30,   Area: 30
3- Perimeter: 32,   Area: 32
4- Perimeter: 36,   Area: 48
5- Perimeter: 48,   Area: 96
6- Perimeter: 60, Area: 120

Area and Perimeter of Trapezoid

1- Perimeter: 40,   Area: 72
2- Perimeter: 27,   Area: 40
3- Perimeter: 41,   Area: 62
4- Perimeter: 38,   Area: 80
5- Perimeter: 44,   Area: 104
6- Perimeter: 60,   Area: 192

Area and Perimeter of Parallelogram

1- Perimeter: $26m$, Area: $30(m)^2$
2- Perimeter: $58m$, Area: $120(m)^2$
3- Perimeter: $44in$, Area: $60(in)^2$
4- Perimeter: $37cm$, Area: $50(cm)^2$
5- Perimeter: $80m$,   Area: $364(m)^2$
6- Perimeter: $60m$, Area: $225(m)^2$

Circumference and Area of Circle

1) Circumference: 43.96 mm   Area: $153.86(mm)^2$
2) Circumference: 21.98 in   Area: $38.465(in)^2$
3) Circumference: 13.816 m   Area: $15.197(m)^2$
4) Circumference: 31.4 cm   Area: $78.5(cm)^2$
5) Circumference: 18.84 in   Area: $28.26(in)^2$
6) Circumference: 28.26 km   Area: $63.59(km)^2$

Perimeter of Polygon

1) 55 mm
2) 42 m
3) 65 cm
4) 28 in
5) 40 m
6) 24 ft

Volume of Cubes

1) $343 m^3$
2) $1,331 (mm)^3$
3) $512 in^3$
4) $3.375 (cm)^3$

5) $2,744(ft)^3$ 6) $12.167(cm)^3$

Volume of Rectangle Prism

1) $560(cm)^3$ 3) $50.4(m)^3$ 5) $140(mm)^3$

2) $57.75(yd)^3$ 4) $288(in)^3$ 6) $1.2(in)^3$

Volume of Cylinder

1) $125.6(cm)^3$ 3) $602.88(yd)^3$ 5) $141.3(m)^3$

2) $42.39(mm)^3$ 4) $510.25(m)^3$ 6) $1,692.46(in)^3$

Volume of Spheres

1) $523.33(in)^3$ 3) $1,436.02(in)^3$ 5) $381.51(in)^3$

2) $33.49(in)^3$ 4) $65.42(in)^3$ 6) $1,071.79(in)^3$

Volume of Pyramid and Cone

1) $326.67\ (in)^3$ 3) $1,050\ (in)^3$ 5) $7.15\ (in)^3$

2) $736.85\ (in)^3$ 4) $183.16\ (in)^3$ 6) $47.1(in)^3$

Surface Area Cubes

1) $1,014(in)^2$ 3) $337.5(in)^2$ 5) $37.5(in)^2$

2) $216(in)^2$ 4) $108(in)^2$ 6) $1,536(in)^2$

Surface Area Rectangle Prism

1) $126(in)^2$ 3) $256(in)^2$ 5) $135.5(in)^2$

2) $149.5(in)^2$ 4) $662(in)^2$ 6) $746(in)^2$

Surface Area Cylinder

1) $87.92(in)^2$ 3) $596.6(in)^2$ 5) $1,381.6(in)^2$

2) $276.32(in)^2$ 4) $62.8(in)^2$ 6) $285.74(in)^2$

Chapter 10: Statistics and probability

Mean, Median, Mode, and Range of the Given Data

Find the mean, median, mode(s), and range of the following data.

1) 24, 59, 20, 37, 14, 24, 47

Mean: __, Median: __, Mode: __, Range: __

2) 6, 13, 13, 19, 15, 10

Mean: __, Median: __, Mode: __, Range: __

3) 21, 35, 49, 11, 45, 27, 35, 19, 14

Mean: __, Median: __, Mode: __, Range: __

4) 25, 11, 1, 15, 25, 18

Mean: __, Median: __, Mode: __, Range: __

5) 24, 14, 14, 17, 23, 15, 14, 29, 29, 8

Mean: __, Median: __, Mode: __, Range: __

6) 7, 14, 19, 11, 8, 19, 8, 15

Mean: __, Median: __, Mode: __, Range: __

7) 29, 28, 66, 76, 14, 44, 18, 44, 22, 44

Mean: __, Median: __, Mode: __, Range: __

8) 35, 35, 57, 78, 59

Mean: __, Median: __, Mode: __, Range: __

9) 16, 16, 29, 46, 54

Mean: __, Median: __, Mode: __, Range: __

10) 13, 9, 3, 3, 5, 6, 7

Mean: __, Median: __, Mode: __, Range: __

11) 4, 12, 4, 6, 1, 8

Mean: __, Median: __, Mode: __, Range: __

12) 8, 9, 15, 15, 17, 17, 17

Mean: __, Median: __, Mode: __, Range: __

13) 7, 7, 1, 16, 1, 7, 19

Mean: __, Median: __, Mode: __, Range: __

14) 13, 17, 10, 12, 12, 18, 15, 19

Mean: __, Median: __, Mode: __, Range: __

15) 9, 14, 19, 19, 29

Mean: __, Median: __, Mode: __, Range: __

16) 6, 6, 16, 18, 15, 22, 37

Mean: __, Median: __, Mode: __, Range: __

17) 25, 11, 14, 25, 18, 13, 7, 5

Mean: __, Median: __, Mode: __, Range: __

18) 55, 34, 34, 48, 85, 7

Mean: __, Median: __, Mode: __, Range: __

19) 54, 28, 28, 65, 5, 8

Mean: __, Median: __, Mode: __, Range: __

20) 88, 84, 23, 26, 11, 88, 19

Mean: __, Median: __, Mode: __, Range: __

Box and Whisker Plot

1) Draw a box and whisker plot for the data set:

   16, 11, 14, 12, 14, 12, 16, 16, 20

2) The box-and-whisker plot below represents the math test scores of 20 students.

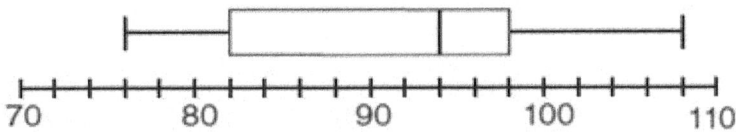

   A. What percentage of the test scores are less than 82?

   B. Which interval contains exactly 50% of the grades?

   C. What is the range of the data?

   D. What do the scores 76, 94, and 108 represent?

   E. What is the value of the lower and the upper quartile?

   F. What is the median score?

Bar Graph

Each student in class selected two games that they would like to play. Graph the given information as a bar graph and answer the questions below:

Game	Votes
Football	13
Volleyball	10
Basketball	18
Baseball	17
Tennis	13

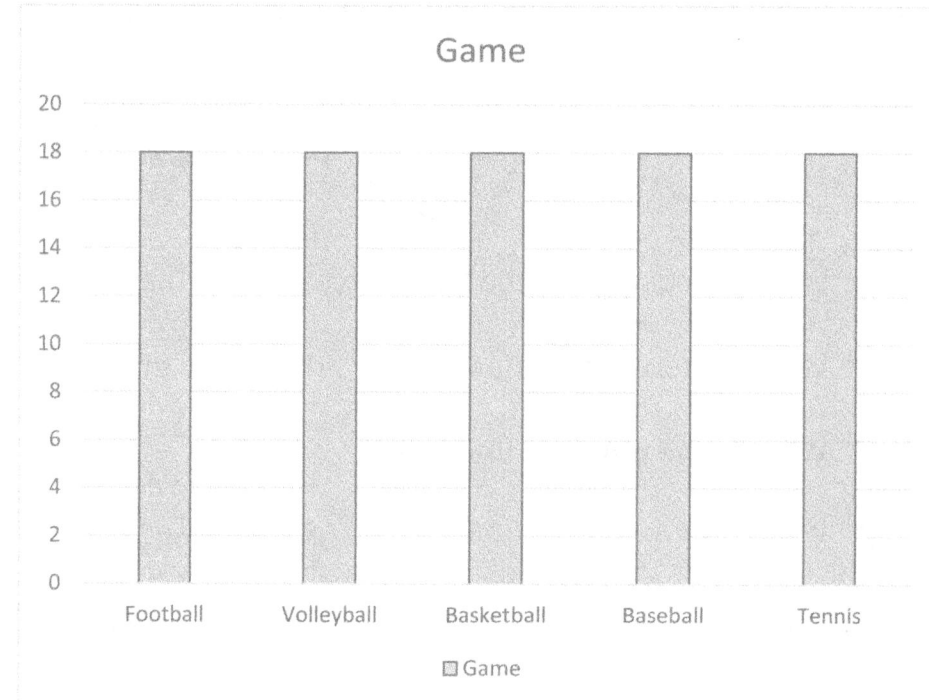

1) Which was the most popular game to play?

2) How many more student like Basketball than Volleyball?

3) Which two game got the same number of votes?

4) How many Volleyball and Football did student vote in all?

5) Did more student like football or Volleyball?

6) Which game did the fewest student like?

Histogram

Create a histogram for the set of data.

Math Test Score out of 100 points.

58	74	63	80	83	65	70	86	67	54
81	73	82	75	71	56	87	66	74	72
84	55	76	73	67	85	69	68	52	87

Frequency Table	
Interval	Number of Values

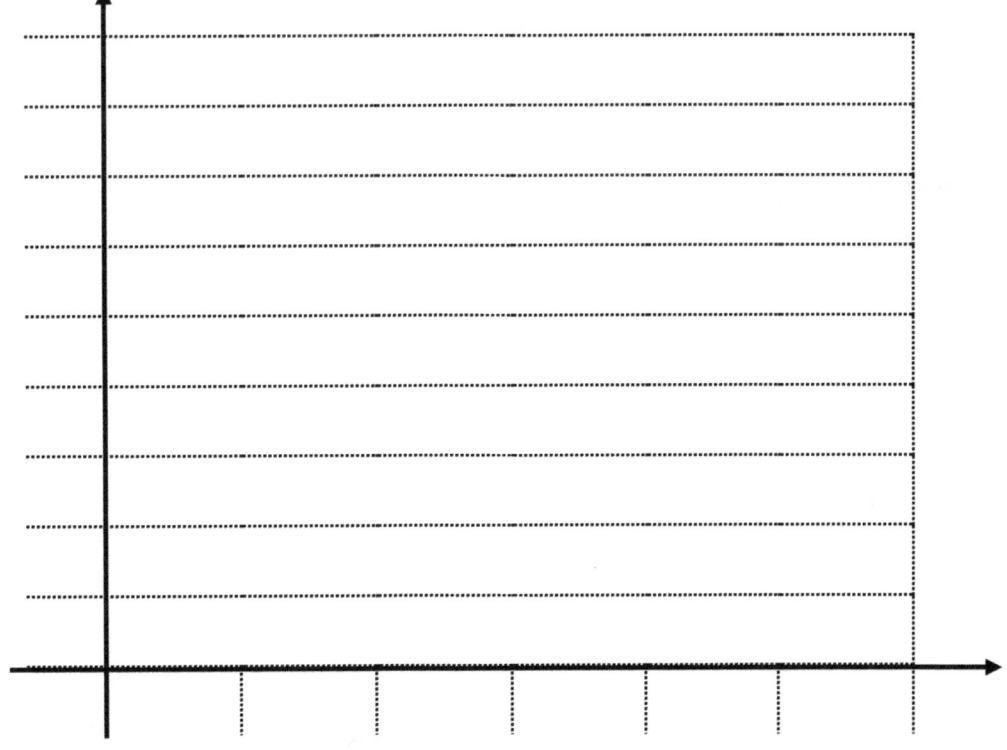

Dot plots

The ages of students in a Math class are given below.

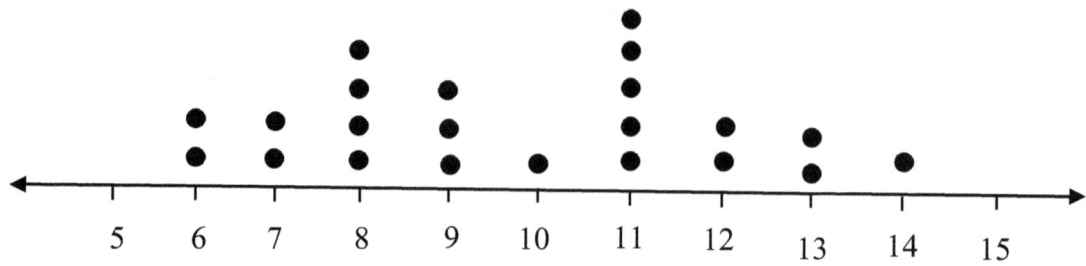

1) What is the total number of students in math class?

2) How many students are at least 12 years old?

3) Which age(s) has the most students?

4) Which age(s) has the fewest student?

5) Determine the median of the data.

6) Determine the range of the data.

7) Determine the mode of the data.

Scatter Plots

A person charges an hourly rate for his services based on the number of hours a job takes.

Hours	Rate
1	$24.5
2	$22
3	$21
4	$19.50

Hours	Rate
5	$19
6	$17.50
7	$17
8	$16.5

1) Draw a scatter plot for this data.

2) Does the data have positive or negative correlation?

3) Sketch the line that best fits the data.

4) Find the slope of the line.

5) Write the equation of the line using slope-intercept form.

6) Using your prediction equation: If a job takes 10 hours, what would be the hourly rate?

Pie Graph

60 people were survey on their favorite ice cream. The pie graph is made according to their responses. Answer following questions based on the Pie graph.

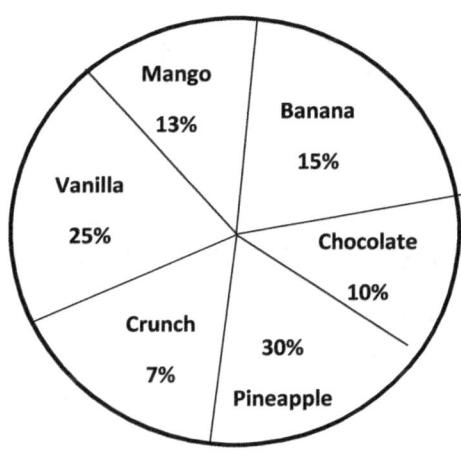

1) How many people like to eat Banana ice cream? _____

2) Approximately, which two ice creams did about half the people like the best? _____

3) How many people said either mango or crunch ice cream was their favorite? _____

4) How many people would like to have chocolate ice cream? _____

5) Which ice cream is the favorite choice of 15 people? _____

Probability

1) A jar contains 16 caramels, 5 mints and 19 dark chocolates. What is the probability of selecting a mint?

2) If you were to roll the dice one time what is the probability it will NOT land on a 4?

3) A die has sides are numbered 1 to 6. If the cube is thrown once, what is the probability of rolling a 5?

4) The sides of number cube have the numbers 4, 6, 8, 4, 6, and 8. If the cube is thrown once, what is the probability of rolling a 6?

5) Your friend asks you to think of a number from ten to twenty. What is the probability that his number will be 15?

6) A person has 8 coins in their pocket. 2 dime, 3 pennies, 2 quarter, and a nickel. If a person randomly picks one coin out of their pocket. What would the probability be that they get a penny?

7) What is the probability of drawing an odd numbered card from a standard deck of shuffled cards (Ace is one)?

8) 32 students apply to go on a school trip. Three students are selected at random. what is the probability of selecting 4 students?

Answer key Chapter 10

Mean, Median, Mode, and Range of the Given Data

1) mean: 32.14, median: 24, mode: 24, range: 45
2) mean: 12.67, median: 13, mode: 13, range: 13
3) mean: 28.44, median: 27, mode: 35, range: 38
4) mean: 15.83, median: 16.5, mode: 25, range: 24
5) mean: 18.7, median: 16, mode: 14, range: 21
6) mean: 12.63, median: 12.5, mode: 19, 8, range: 12
7) mean: 38.5, median: 36.5, mode: 44, range: 62
8) mean: 52.8, median: 57, mode: 35, range: 43
9) mean: 32.2, median: 29, mode: 16, range: 38
10) mean: 6.57, median: 6, mode: 3, range: 10
11) mean: 5.83, median: 5, mode: 4, range: 11
12) mean: 14, median: 15, mode: 17, range: 9
13) mean: 8.29, median: 7, mode: 7, range: 18
14) mean: 14.5, median: 14, mode: 12, range: 9
15) mean: 18, median: 19, mode: 19, range: 20
16) mean: 17.14, median: 16, mode: 6, range: 31
17) mean: 14.75, median: 13.5, mode: 25, range: 20
18) mean: 43.83, median: 41, mode: 34, range: 78
19) mean: 31.33, median: 28, mode: 28, range: 60
20) mean: 48.43, median: 26, mode: 88, range: 77

Box and Whisker Plot

1)

2)

THEA Math Practice Book

A. 25%

B. 94

C. 32

D. Minimum, Median, and Maximum

E. Lower (Q_1) is 82 and upper (Q_3) is 98

F. 94

Bar Graph

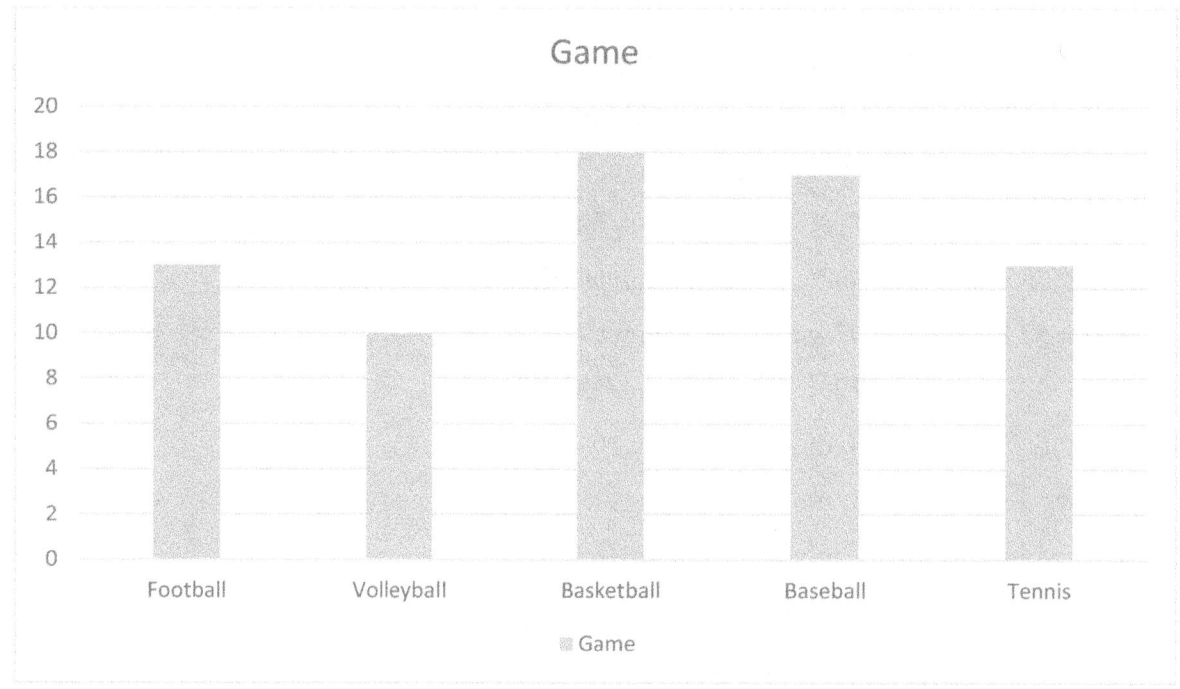

1) Basketball
2) 8 students
3) Football and Tennis
4) 23
5) Football
6) Volleyball

Histogram

Frequency Table	
Interval	Number of Values
52-57	4
58-63	2
64-69	6
70-75	8
76-81	3
82-87	7

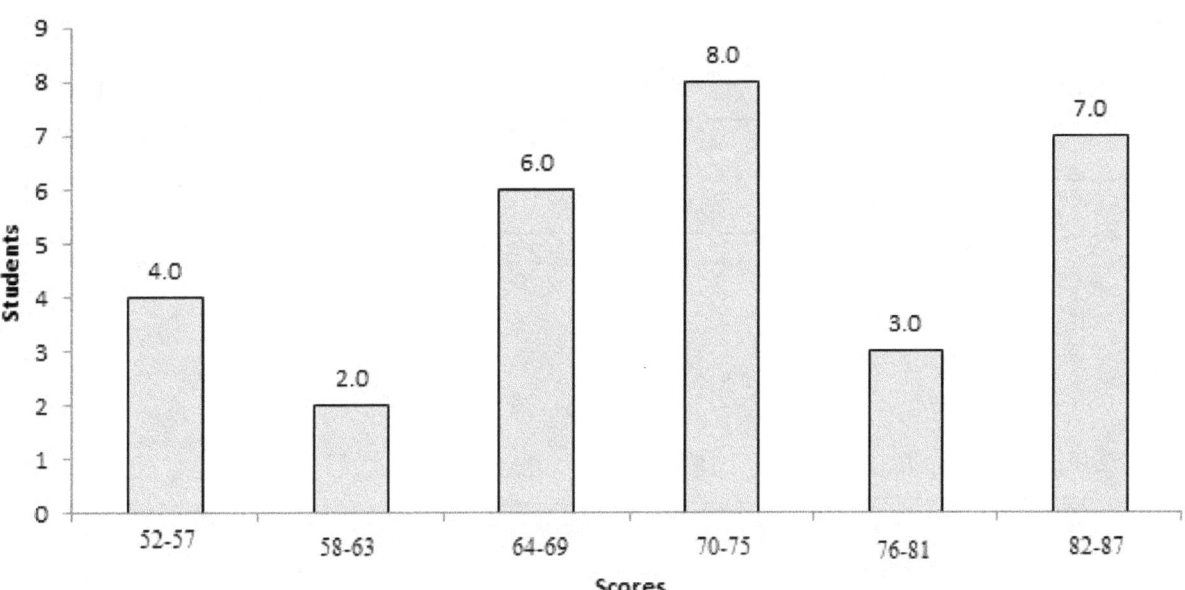

Dot plots

1) 22
2) 5
3) 11
4) 10 and 14
5) 2
6) 4
7) 2

Scatter Plots

1)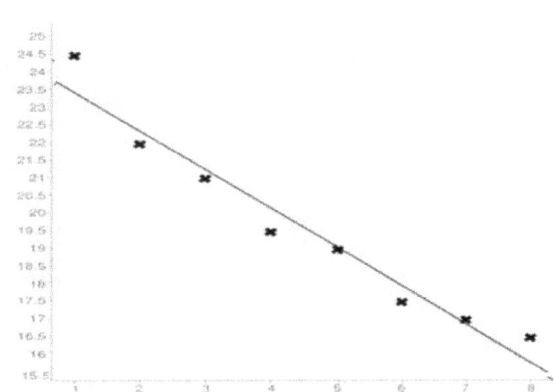

2) Negative correlation
3) ----
4) Slope(m)= -1
5) $y = -x + 24.5$
6) 14.5

THEA Math Practice Book

Stem–And–Leaf Plot

1)

Stem	leaf
1	2 5 6 8 9
3	1 2 4 7 8 9
5	4 5

2)

Stem	leaf
1	4 7
4	1 2 4 6 8 9
7	2 2 4 8

3)

Stem	leaf
6	2 5 5 6 8
10	5 8 9
12	4 5 6 7

4)

Stem	leaf
4	2 9 5
6	0 1 3 3 6 8
9	0 7 9

5)

Stem	leaf
1	5
5	5 6 8
10	2 5 8

6)

Stem	leaf
5	5 7 8
7	1 3 7 9
12	0 3 3 4 7

Pie Graph

1) 9

2) Vanilla and pineapple

3) 12

4) 6

5) Vanilla

Probability

1) $\frac{1}{8}$

2) $\frac{5}{6}$

3) $\frac{1}{6}$

4) $\frac{1}{3}$

5) $\frac{1}{10}$

6) $\frac{3}{8}$

7) $\frac{5}{13}$

8) $\frac{1}{8}$

THEA Test Review

THEA Test Mathematics Formula Sheet

Area of a:

Parallelogram $\qquad A = bh$

Trapezoid $\qquad A = \frac{1}{2}h(b_1 + b_2)$

Surface Area and Volume of a:

Rectangular/Right Prism $\qquad SA = ph + 2B \qquad V = Bh$

Cylinder $\qquad SA = 2\pi rh + 2\pi r^2 \qquad V = \pi r^2 h$

Pyramid $\qquad SA = \frac{1}{2}ps + B \qquad V = \frac{1}{3}Bh$

Cone $\qquad SA = \pi rs + \pi r^2 \qquad V = \frac{1}{3}\pi r^2 h$

Sphere $\qquad SA = 4\pi r^2 \qquad V = \frac{4}{3}\pi r^3$

(p = perimeter of base B; $\pi = 3.14$)

Algebra

Slope of a line $\qquad m = \dfrac{y_2 - y_1}{x_2 - x_1}$

Slope-intercept form of the equation of a line $\qquad y = mx + b$

Point-slope form of the Equation of a line $\qquad y - y_1 = m(x - x_1)$

Standard form of a Quadratic equation $\qquad y = ax^2 + bx + c$

Quadratic formula $\qquad x = \dfrac{-b \pm \sqrt{b^2 - 4ac}}{2a}$

Pythagorean theorem $\qquad a^2 + b^2 = c^2$

Simple interest

$$I = prt$$
(I = interest, p = principal, r = rate, t = time)

THEA Practice Test 1

Mathematics

Total Number of Questions: 50 Questions

Total time: 240 Minutes (All three sections)

You may use a non-programmable calculator for this test.

Administered *Month Year*

THEA Math Practice Book

1) How many odd integers are between $\frac{-69}{6}$ and $\frac{19}{5}$?

    A. 5

    B. 11

    C. 8

    D. 9

2) Which expression correctly represents the distance between the two points shown on the number line?

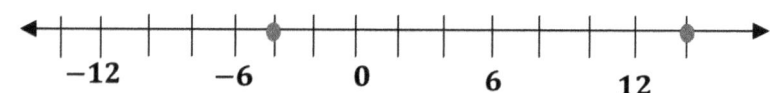

    A. $-4 - 14$

    B. $|-4 + 14|$

    C. $-4 + 14$

    D. $|4 + 14|$

3) In the triangle ABC, if angle B and angle C both equal 60°, then what is the length of side BC?

    A. 3 cm

    B. 6 cm

    C. 12 cm

    D. 4 cm

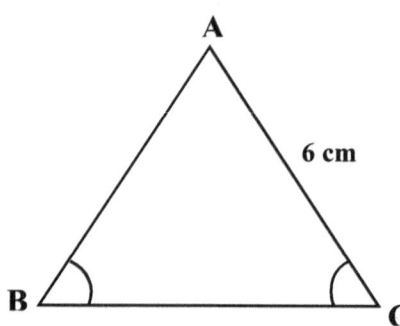

4) Which of the following represents the sum of the factors of 21?

   A. 32

   B. 29

   C. 11

   D. 23

5) Arrange the following fractions in order from least to greatest.

$$\frac{3}{8}, \frac{4}{7}, \frac{1}{5}, \frac{23}{25}, \frac{14}{19}$$

   A. $\frac{1}{5}, \frac{3}{8}, \frac{4}{7}, \frac{14}{19}, \frac{23}{25}$

   B. $\frac{3}{8}, \frac{4}{7}, \frac{1}{5}, \frac{14}{19}, \frac{23}{25}$

   C. $\frac{23}{25}, \frac{14}{19}, \frac{3}{8}, \frac{4}{7}, \frac{1}{5}$

   D. $\frac{14}{19}, \frac{23}{25}, \frac{1}{5}, \frac{4}{7}, \frac{3}{8}$

6) Elena earns $7.00 an hour and worked 33 hours. Her brother earns $8.25 an hour. How many hours would her brother need to work to equal Elena's earnings over 36 hours?

   A. 22.25

   B. 28

   C. 40

   D. 30.5

7) In a library, 30% of the books are fiction and the rest are non-fiction. Given that there are 1,200 more non-fiction books than fiction books, what is the total number of books in the library?

   A. 2,000

   B. 2,800

   C. 3,000

   D. 3,500

8) A map has the scale of 5 cm to 1 km. What is the actual area of a lake on ground which is represented as an area of 85 cm^2 on the map?

   A. 4.2 km^2

   B. 3.4 km^2

   C. 15 cm^2

   D. 1.5 cm^2

9) $45.45 \div 0.9 =$

   A. 50.05

   B. 50.50

   C. 5.05

   D. 5.50

10) What is the circumference of a circle with a radius of 11 inches?

   A. 11π

   B. 22π

   C. 121π

   D. 30.25π

11) Alfred needs to calculate his monthly water bill. His family used 28,500 gallons at a rate of $0.72 per hundred gallons. Also, there is a monthly fee of $4.40 on each period. What is his total bill?

   A. $2,310.3

   B. $205.6

   C. $209.6

   D. $29.6

12) Simplify $\frac{(2x^4+6x^3)}{(x^3+3x^2)} = ?$

   A. $2x$

   B. $x - 3$

   C. $3x$

   D. $3x(x+2)$

13) Which of the following expressions is undefined in the set of real numbers?

   A. $\sqrt[2]{144}$

   B. $\sqrt[3]{-27}$

   C. $\sqrt{-49}$

   D. $\sqrt[4]{254}$

14) If $f(x) = 3x^2$, and 5f(2a) = 180 then what could be the value of a?

   A. -3

   B. -6

   C. 6

   D. 3

15) Speed of a train is 160 miles per hour for 4 hours and 45 minutes. How many miles did the train travel?

   A. 712

   B. 625

   C. 725

   D. 655

16) Solve $9x - 7 \leq 15x + 11$

   A. $x \leq -3$

   B. $x \geq -3$

   C. $x \leq 3$

   D. $x \geq 3$

17) What is the value of 3^6?

   A. $(2+2)^8$

   B. 9^3

   C. $3(3^2)$

   D. $3^3 + 3^3$

18) Shane Williams puts $3,600 into a saving bank account that pays simple interest of 4.5%. How much interest will she earn after 2 years?

   A. $2,845

   B. $ 2,324

   C. $324

   D. $146

19) Three angles join to form a straight angle. One angle measure 72°. Other angle measures 34°. What is the measure of third angle?

   A. 18°

   B. 68°

   C. 106°

   D. 74°

THEA Math Practice Book

20) Evaluate $\frac{21x^9 y^2 z^{-3}}{14 x^5 y^6 z^3}$.

    A. $\frac{3x^4 y^3}{2 z^2}$

    B. $\frac{3x^3 z^2}{2 y^2}$

    C. $\frac{3x^4}{2 y^4 z^6}$

    D. $\frac{7y^3}{2 x^2 z^3}$

21) Find the equation for line passing through $(2, -4)$ and $(5, 1)$.

    A. $-2x - 5y = 22$

    B. $3y - 5x = -22$

    C. $-3x + 10y = 12$

    D. $5y + 3x = -22$

22) If $x = -5$ and $y = 7$, calculate the value of $\frac{x^2+7}{y-3}$.

    A. $\frac{1}{8}$

    B. -8

    C. 8

    D. 4

THEA Math Practice Book

23) The table shows the parking rates for the outside terminal area at an airport.

Ethan parked at the lot for $3\frac{1}{2}$ hours. How much did he owe?

A. $8.65

B. $7.65

C. $7.7

D. $7.85

2 hours	$3.20
Each 30 minutes after 2 hours	$1.5
24-hours Discount rate	$40

24) What is the value of x in term of c and d $\frac{c-d}{dx} = \frac{2}{7}$? (c and d>0)

A. $\frac{7}{2}(\frac{c}{d} - 1)$

B. $\frac{7}{2}(1 - \frac{c}{d})$

C. $\frac{2}{7}(\frac{c}{d} - 1)$

D. $\frac{2}{7}(1 - \frac{c}{d})$

25) Factor the equation $9x^4 - 4x^2$.

A. $2x^2(3x + 2)$

B. $x^2(2x + 3)(x - 3)$

C. $x^2(3x - 1)(x + 3)$

D. $x^2(3x - 2)(3x + 2)$

26) Which of the following are the solutions to the equation $x^2 - 14x + 48 = 0$?

   A. $-6, 8$

   B. $8, 6$

   C. $-8, -6$

   D. $6, -8$

27) Which of the following equations best represents the line in the graph below?

   A. $y = \frac{1}{5}x + 4$

   B. $y = x + 5$

   C. $y = \frac{1}{5}x - 5$

   D. $y = 5x + 1$

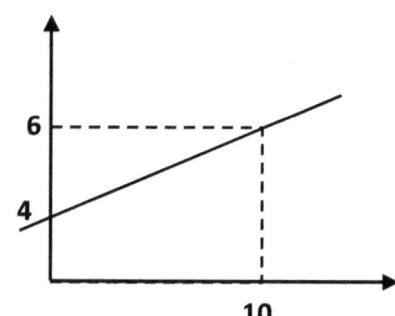

28) If $-3x + 2y = -2$ and $4x - 3y = 5$, what is the value of x?

   A. 7

   B. 1

   C. -2

   D. -4

29) lengths of two sides of a triangle are 8 and 5. Which of the following could Not be the measure of third side?

   A. 2

   B. 4

   C. 10

   D. 12

30) The following data set is given: 111, 140, 127, 165, 159, 120.

Adding which number to the set will increase its mean?

A. 134

B. 129

C. 132

D. 148

31) An award for best education improvement is awarded annually to a winning US state, and the winners from 2002 to 2011 are given in the table below. Find the mode of this set of states.

Year	2002	2003	2004	2005	2006	2007	2008	2009	2010	2011
State	New Jersey	New York	Oregon	New York	Oregon	California	Oregon	Ohio	Oregon	New York

A. Ohio and Oregon

B. California and Ohio

C. New York and Oregon

D. New Jersey and California

32) How long is a distance of 8 km if measured on a map with a scale of 1:50,000?

A. 14

B. 16

C. 8

D. 3

33) What is the area of a square, if its side measures $\sqrt{13}$ m?

   A. $\sqrt{13}$

   B. $3\sqrt{13}$

   C. $4\sqrt{13}$

   D. 13

34) What is the answer of the following equation $2x^4y^4 (3x^3y^3)^4 =$?

   A. $81x^{11}y^{11}$

   B. $162x^{16}y^{16}$

   C. $46x^{10}y^8$

   D. $112x^8y^8$

35) The circle graph below shows the type of pizza that people prefer for lunch. If 220 people were surveyed, how many people preferred Margherita?

   A. 65

   B. 50

   C. 55

   D. 20

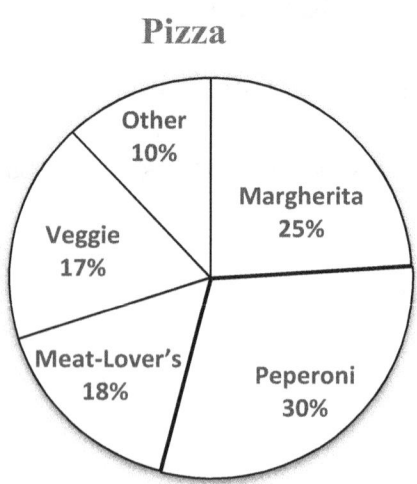

36) Rosie is *x* years old. She is 3 years older than her twin brothers Milan and Marcel. What is the mean age of the three children?

   A. $x + 2$

   B. $x - 3$

   C. $x + 3$

   D. $x - 2$

37) a is inversely proportional to $(2b - 5)$. If a = 8 and b = 9, express a in terms of b.

   A. $a = 2b - 18$

   B. $a = 108(2b - 5)$

   C. $a = \frac{104}{2b-5}$

   D. $a = \frac{2b-5}{104}$

38) A baseball has a volume of 36π. What is the length of the diameter?

   A. 15

   B. 6

   C. 3

   D. 12

39) 154 is What percent of 140?

   A. 110 %

   B. 60 %

   C. 120 %

   D. 220 %

40) A phone manufacturer makes 18,000 phone a year. The company randomly selects 600 of the phones to sample for inspection. The company discovers that there are 5 faulty phones in the sample. Based on the sample, how many of the 18,000 total phones are likely to be faulty?

   A. 300

   B. 50

   C. 250

   D. 150

41) Emma and Mia buy a total of 28 books. Emma bought 4 more books than Mia did. How many books did Emma buy?

   A. 24

   B. 16

   C. 12

   D. 8

42) A bag contains 11 white balls, 6 red balls and 7 black balls. A ball is picked from the bag at random. Find the probability of picking a red ball?

   A. 1.25

   B. 2.25

   C. 0.45

   D. 0.25

43) The radius of the following cylinder is 4 inches, and its height are 9 inches.

What is the surface area of the cylinder in square inches? (π=3.14)

A. 326.56

B. 366.5

C. 5.56

D. 36.65

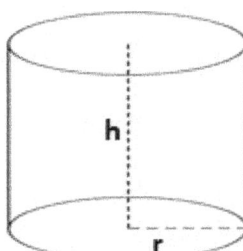

44) The line n has a slope of $\frac{a}{b}$, where a and b are integers. What is the slope of a line that is perpendicular to line n?

A. $-\frac{a}{b}$

B. $\frac{a}{b}$

C. $\frac{b}{a}$

D. $-\frac{b}{a}$

45) In a store, 38% of customers are female. If the total number of customers is 750, then how many male customers are dealing with the store?

A. 421

B. 726

C. 465

D. 356

46) If 66.33 kg is divided into two parts, in a ratio of 8:3, how many kg is the smaller share?

   A. 48.24 kg

   B. 18.09 kg

   C. 1.89 kg

   D. 19.09 kg

47) Express as a single fraction in its simplest form: $\frac{5}{(x+2)} - \frac{2}{(4x-1)} = ?$

   A. $\frac{18x-9}{(x+2)(4x-1)}$

   B. $\frac{9x-18}{(x+2)(4x-1)}$

   C. $\frac{15x-17}{4x+2}$

   D. $\frac{-17}{4x-1}$

48) Ryan is x years old, and her sister Mitzi is $(6x - 20)$ years old. Given that Mitzi is twice as old as Ryan, what is Mitzi's age?

   A. 7

   B. 11

   C. 10

   D. 14

49) The set of possible values of p is {3, 5, 11}. What is the set of possible values of h if $2h = 5p + 1$?

A. {6,17,25}

B. {8,14,26}

C. {8, 13, 28}

D. {6,11,28}

50) In the infinitely repeating decimal below, 4 is the second digit in the repeating pattern. What is the 764th digit? $\frac{1}{21} = \overline{0.047619}$

A. 4

B. 7

C. 9

D. 0

"End of THEA Practice Test 1."

THEA Practice Test 2

Mathematics

Total Number of Questions: 50 Questions

Total time: 240 Minutes (All three sections)

You may use a non-programmable calculator for this test.

Administered *Month Year*

1) What is the sum of the smallest prime number and five times the largest negative even integer?

   A. -10

   B. 14

   C. -8

   D. -12

2) Sam's incomes and expenditures for the first season of the last year are given in the table below. In which month were her savings the highest?

   A. January

   B. March

   C. February

   D. All months were the same.

Month	Income	Cost
January	$3,740	$2,202
February	$3,855	$1,985
March	$3,970	$2,005

3) What is the number a, if the result of adding a to 29 is the same as subtracting $4a$ from 244?

   A. -59

   B. 34

   C. 122

   D. 43

4) Solve these fractions and reduce to its simplest terms: $8\frac{7}{24} - 6\frac{2}{3} + 2\frac{5}{6} =$

   A. $4\frac{11}{24}$

   B. $-4\frac{7}{24}$

   C. $\frac{7}{8}$

   D. $5\frac{3}{8}$

5) Calculate, $3^3 + 3 + 3^0 = ?$

   A. 3^4

   B. 30

   C. 31

   D. 35

6) Which decimal is equivalent to $\frac{153}{170}$?

   A. 0.109

   B. 1.09

   C. 0.09

   D. 0.9

7) The price of a shirt increased from $30 to $32.10. What is the percentage increase in the price?

   A. 7%

   B. 0.7%

   C. 0.93%

   D. 1.7%

8) Find the solution set of the following equation: $|2x - 3| = 7$

   A. $\{5, -2\}$

   B. $\{2, -5\}$

   C. $\{2\}$

   D. $\{-5, -2\}$

9) What is the solution to the pair of equations below? $\begin{cases} 2x - 5y = 16 \\ 2x + y = 4 \end{cases}$

   A. $x = 0$ and $y = -3$

   B. $x = -3$ and $y = 0$

   C. $x = 3$ and $y = -2$

   D. $x = 3$ and $y = -3$

10) The train each 30 minutes passes an average 5 stations. At this rate, how many stations will it pass in three hours.

   A. 20

   B. 30

   C. 45

   D. 60

11) If Sofia buy a shirt marked down 16 percent from its $160 while Natalia buys the same shirt mark down only 11 percent, how much more does Natalia pay for the shirt?

A. $80

B. $4.8

C. $16

D. $8

12) Find the circumference of the circle in terms of π.

A. 9π in.

B. 72π in.

C. 36 π in.

D. 112π in.

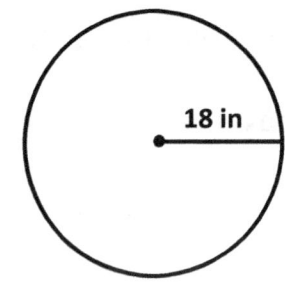

13) Find the volume of rectangular prism below?

A. 1,880 in^3

B. 1,080 in^3

C. 2,280 in^3

D. 285 in^3

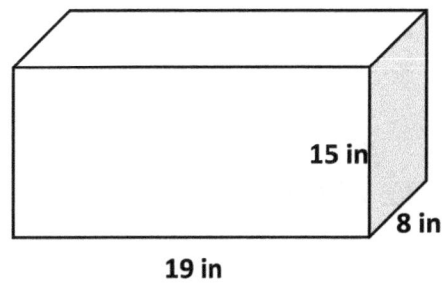

14) What is $\sqrt[4]{5^{-8}}$ in simplest form?

A. $\dfrac{1}{2,250}$

B. $\dfrac{1}{125}$

C. $\dfrac{1}{5}$

D. $\dfrac{1}{25}$

15) Which statement correctly describes the value of N in the equation below?

$7(4N - 10) = 4(7N - 15)$?

A. N has no correct solutions.

B. N=1 is one solution.

C. N has infinitely many correct solutions.

D. N=0 is one solution.

16) What is the maximum amount of grain, the silo can hold, in cubic feet?

A. $1,620\pi\ m^3$

B. $405\pi\ m^3$

C. $810\pi\ m^3$

D. $1,200\pi\ m^3$

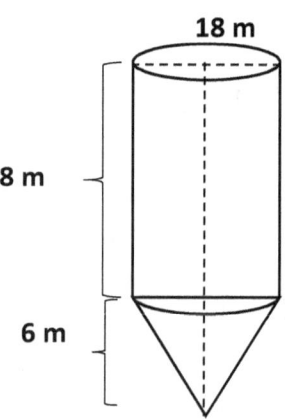

17) What is 7.25×10^{-5} in standard form?

   A. −72,500

   B. −0.0000725

   C. $\frac{1}{725,000}$

   D. 0.0000725

18) What is the value of x in the triangle?

   A. 64°

   B. 130°

   C. 50°

   D. 60°

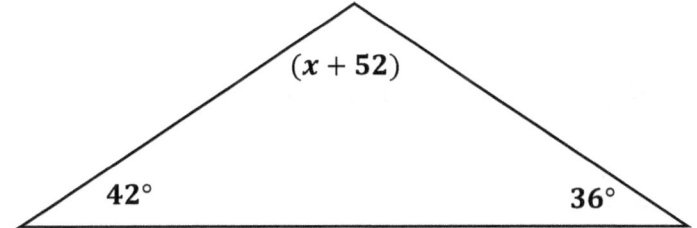

19) Find the length of the unknown side.

   A. 22 ft

   B. 20 ft

   C. 18 ft

   D. 12 ft

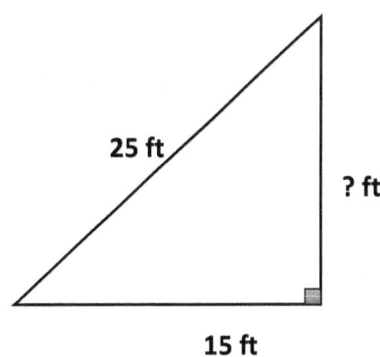

20) A store sells all of its products at a price 12% greater than the price the store paid for the product. How much does the store sell a product if the store paid $150 for it?

A. $180

B. $160

C. $168

D. $38

21) What is the area of shaded region?

A. 95

B. 35

C. 130

D. 165

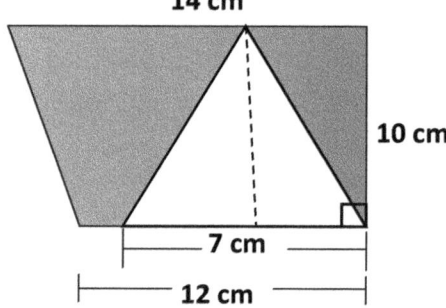

22) if $xy - 8x = 54$ and $y - 8 = 6$, then $x =$?

A. 8

B. 14

C. 6

D. 9

23) Which is the value of x in the equation $\frac{x}{4} = x - 6$?

    A. 4

    B. 6

    C. 8

    D. 12

24) What is the value of x, If $-5x + 4y = 7$ and $-4x + 3y = 5$?

    A. -2

    B. 0

    C. 3

    D. 1

25) What is the probability of Not spinning at F?

    A. $\frac{5}{8}$

    B. $\frac{3}{8}$

    C. $\frac{1}{8}$

    D. $\frac{7}{8}$

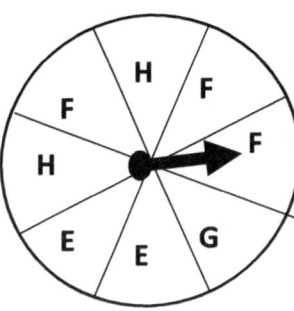

26) Ella bought 32 movies for $5.04 per movie. Which equation shows the BEST estimate of the total cost?

    A. 30 × $6 = $180

    B. 30 × $5 = $150

    C. 32 × $5 = $160

    D. 32 × $6 = $192

27) A pick-up truck travels 50 mile on 7 L of gasoline when driven on a smooth road. If the cost of gasoline is $1.40/L, which is the cost of 1,500 mile of highway (smooth)?

   A. $141

   B. $1,249

   C. $129

   D. $294

28) A position of subway station and Grace's house shown by a grid. The station is located at $(-6, -9)$, and her house is located at $(2, -3)$. What is the distance between her house and the subway stop?

   A. 6

   B. 10

   C. $10\sqrt{2}$

   D. 15

29) For the following set of numbers find the median.

   225, 75, 280, 89, 198, 124, 512.

   A. 198

   B. 124

   C. 225

   D. 89

30) What is solution to the equation $\sqrt{4x-7} = 9$?

   A. -18

   B. -14

   C. 10

   D. 22

31) The equation $x = 4y - 8$ has a y-intercept of?

   A. 2

   B. -2

   C. $\frac{1}{2}$

   D. $-\frac{1}{2}$

32) If $3^{2x} = 729$, then $x = ?$

   A. 4

   B. 6

   C. 3

   D. 2

33) Which is the smallest positive integer which is divisible by both 18 and 64.

   A. 108

   B. 54

   C. 576

   D. 124

34) A cube has total surface area of 96 cm². what is the volume of the cube in cm³?

   A. 8

   B. 64

   C. 36

   D. 108

35) Which is the value of x^2, if $x^2 + 2x = 35$?

   A. 12

   B. -7

   C. 36

   D. 25

36) Each of 5 pitchers can contain up to $\frac{3}{5}$ L of water. If each of the pitcher is at least the half full, which of the following expressions represents the total amount of water, W, contained on all 5 pitchers?

   A. $2.5 < w < 5$

   B. $1.5 < w < 5$

   C. $0.5 < w < 3$

   D. $1.5 < w < 3$

37) Ages of the players on a volleyball team is given. Which is the range of their ages? 47, 32, 17, 65, 22, 40, 28, 57, 20, 18, 56, 39

A. 38

B. 19

C. 48

D. 52

38) Find the area of the circle to the nearest tenth. Use 3.14 for π.

A. 201

B. 202

C. 200.9

D. 109

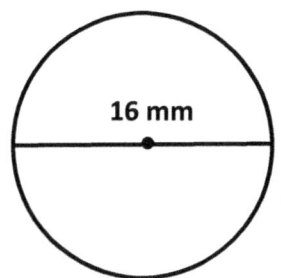

39) What is the simplest form of the expression $\frac{2x^2-15x+7}{4(x^2-\frac{1}{4})}$?

A. $\frac{x+7}{2(x-\frac{1}{2})}$

B. $\frac{x-7}{2x+1}$

C. $\frac{x+7}{2x+1}$

D. $\frac{x-7}{x-2}$

40) What is the value of $\left(\frac{1}{3}\right)^{-4}$?

A. $\frac{1}{81}$

B. -81

C. $-\frac{1}{81}$

D. 81

41) What is the simplest form of the expression $\frac{(5x^{-3}y^4)^3}{100y^{-2}z^2}$, (using positive exponent)?

A. $\frac{4y^{14}z}{59}$

B. $\frac{5y^{14}}{4x^9z^2}$

C. $\frac{5x^8z^2}{4zy^8}$

D. $\frac{x^{14}}{y^9}$

42) Some fruit sells for $30 per kilograms. What is the price in cent per gram?

A. 0.003

B. 0.3

C. 0.03

D. 3

THEA Math Practice Book

43) The price of water triples every 5 years. If the price of water on January 1st, 2012, is $6 per gallon, what is the equation that would be used to calculate the price(P) of water on January 1st, 2007?

   A. $6P = 6$

   B. $\dfrac{P}{6} = 3$

   C. $3p = 6$

   D. $6P = 3$

44) The line $4y + 5 = 20x + 9$ and $5y + 4 = x + 7$ are?

   A. Perpendicular

   B. Parallel

   C. The same line

   D. Neither parallel nor perpendicular

45) A rectangular box measures $5\frac{1}{3}$ feet by $8\frac{2}{5}$ feet. It is divided into four equal parts. What is the area of one of those parts?

   A. 11

   B. 9.2

   C. 11.2

   D. 22.4

46) Find the slope of the line.

A. $\dfrac{1}{2}$

B. $\dfrac{1}{3}$

C. 2

D. $-\dfrac{1}{4}$

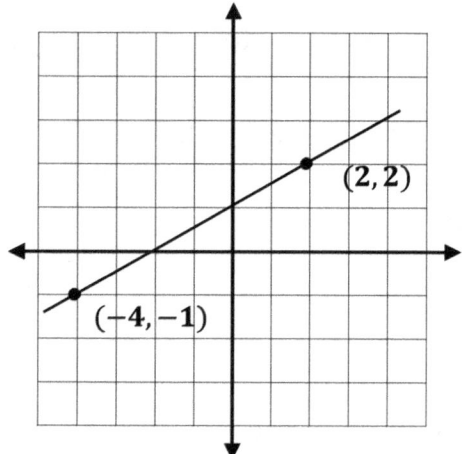

47) Amelia cuts a piece of birthday cake as shown below. What is the volume of the piece of cake?

A. 425 cm^3

B. 375 cm^3

C. 750 cm^3

D. 250 cm^3

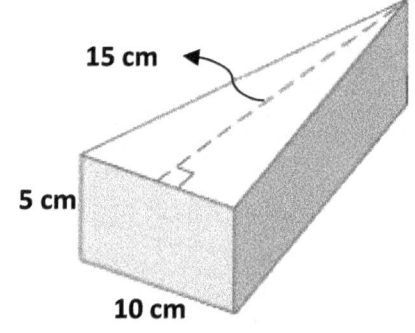

48) Let $f(x) = 4x - 7$. If $f(a) = -19$ and $f(b) = 9$, then what is $f(a + b)$?

A. 3

B. 5

C. -3

D. -5

49) Given that $28x = 12y$, find the ratio $x : y$.

   A. $5 : 9$

   B. $4 : 11$

   C. $6 : 5$

   D. $3 : 7$

50) What is the number of sides of a regular polygon whose interior angles are 168° each? (Remember, the sum of exterior angles of any polygon is 360°).

   A. 6

   B. 30

   C. 22

   D. 12

"End of THEA Practice Test 2."

Answers and Explanations

Answer Key

Now, it's time to review your results to see where you went wrong and what areas you need to improve!

THEA Math Practice Test

Practice Test 1

1	C	20	C	39	A
2	D	21	B	40	D
3	B	22	C	41	B
4	A	23	C	42	D
5	A	24	A	43	A
6	B	25	D	44	D
7	C	26	B	45	C
8	B	27	A	46	B
9	B	28	D	47	A
10	B	29	A	48	C
11	C	30	D	49	C
12	A	31	C	50	A
13	C	32	B		
14	D	33	D		
15	A	34	B		
16	B	35	C		
17	B	36	D		
18	C	37	C		
19	D	38	B		

Practice Test 2

1	C	20	C	39	B
2	B	21	A	40	D
3	D	22	D	41	B
4	A	23	C	42	D
5	C	24	D	43	C
6	D	25	A	44	D
7	A	26	C	45	C
8	A	27	D	46	A
9	C	28	B	47	B
10	B	29	A	48	C
11	D	30	D	49	D
12	C	31	A	50	B
13	C	32	C		
14	D	33	C		
15	A	34	B		
16	C	35	D		
17	D	36	D		
18	C	37	C		
19	B	38	A		

THEA Math Practice Book

THEA Practice Test 1
Answers and Explanations

1) Answer: C

$\frac{-69}{6} = -11.5$ and $\frac{19}{5} = 3.8$, then the odd numbers are:

$(-11, -9, -7, -5, -3, -1, 1, 3)$

2) Answer: D

The distance between two points always is positive. Use formula:

AB = |b − a| or |a − b| → |− 4 − 14| or |14 − (−4)| = |14 + 4|

3) Answer: B

Sum of the measures of the angles of a triangle is 180, if two angles are 60 then the third one is 60, then the triangle is equilateral triangle, and all side are equal.

4) Answer: A

All factors of 21 are: 1, 3, 7, 21, then sum of them is 32.

5) Answer: A

Rewriting each fraction with common denominator or converting each fraction to decimal and order the decimal from least to greatest.

$\frac{3}{8} = 0.375$ $\frac{4}{7} = 0.57$ $\frac{1}{5} = 0.2$ $\frac{23}{25} = 0.92$ $\frac{14}{19} = 0.73$

6) Answer: B

calculating Elena's total earnings: 33 hours × $7.00 an hour = $231

Next, divide this total by her brother's hourly rate:

$231 ÷ $8.25 = 28 hours

7) Answer: C

number of fiction books: x ; number of nonfiction books: $x + 1,200$

Total number of books: $x + (x + 1,200)$

30% of the total number of books are fiction, therefore:

$30\%[x + (x + 1,200)] = x \rightarrow 0.3(2x + 1,200) = x$

WWW.MathNotion.com

$0.6x + 360 = x \to 360 = x - 0.6x \to 0.4x = 360$

$\to x = 900$ number of fictions

$x + 1,200 = 900 + 1,200 = 2,100$, the number of nonfiction books

$900 + 2,100 = 3,000$, the total number of books in the library

8) Answer: B

The scale is: 5 cm:1km, (5 cm on the map represents an actual distance of 1 km).

first necessary to rewrite the scale ratio in terms of units squared:

5^2cm square:1^2km square, which gives $25\ cm^2 : 1\ km^2$

Then, $\frac{25\ cm^2}{1 km^2} = \frac{85\ cm^2}{x\ km^2}$, (where x is the unknown actual area).

Every proportion you write should maintain consistency in the ratios described (km^2 both occupy the denominator).

Cross-multiply and isolate to solve for the unknown area x:

$x \cdot \frac{25\ cm^2}{1 km^2} = 85\ cm^2 \to x = 85\ cm^2 \cdot \frac{1\ km^2}{25\ cm^2} \to x = 3.4\ km^2$

9) Answer: B

move decimal point in divisor so last digit is in the unit place (0.9 to 9)

move decimal point in dividend same number of places to the right,

(45.45 to 454.5); divide (4,545 ÷ 9)=505

insert a decimal point into the answer above the decimal point in the dividend (50.5)

10) Answer: B

Circumference= $2\pi r = 2 \times \pi \times 11 = 22\pi$

11) Answer: C

Be careful with the conversion factor (per hundred gallons; NOT per gallon).

$28,500 \times \frac{0.72}{100} = 205.2$

$205.2 + 4.40 = 209.6$

12) Answer: A

$$\frac{(2x^5+6x^4)}{(x^4+3x^2)} = \frac{2x^4(x+3)}{x^3(x+3)} = 2x$$

13) Answer: C

For odd index we can have negative radicand.

In the even index, negative radicand is undefined.

$\sqrt{-49}$ has a negative number under the even index, so it is non-real.

Negative numbers don't have real square roots, because negative and positive integer squared is either positive or 0.

14) Answer: D

$5f(2a) = 180 \rightarrow$ (divide by 5): $f(2a) = 36$

(subtitute 2a) $3(2a)^2 = 108 \rightarrow$ (divide by3): $(2a)^2 = 36 \rightarrow 4a^2 = 36 \rightarrow$

$a^2 = 9 \rightarrow a = 3$

15) Answer: A

4 hours and 45 minutes is 4.45 hour.

$R \times T = D \rightarrow 160 \times 4.45 = D \rightarrow D = 712$ miles

16) Answer: B

$9x - 7 \leq 15x + 11 \rightarrow Add\ 7: 9x - 7 + 7 \leq 15x + 7 + 11 \rightarrow$

$subtract\ 15x: 9x - 15x \leq 15x - 15x + 18 \rightarrow -6x \leq 18 \rightarrow x \geq -3$

17) Answer: B

Use formula to raise a number: $(x^a)^b = x^{ab} \rightarrow 3^6 = (3^2)^3 = 9^3$

18) Answer: C

Simple interest rate: I = prt (I = interest, p = principal, r = rate, t = time)

$I = 3,600 \times 0.045 \times 2 = 324$

19) Answer: D

A straight angle is an angle measured exactly 180°

$72° + 34° = 106° \rightarrow 180° - 106° = 74°$

THEA Math Practice Book

20) Answer: C

$$\frac{21x^9y^2z^{-3}}{14\,x^5y^6z^3} = \frac{21}{14} \times \frac{x^9}{x^5} \times \frac{y^4}{y^8} \times \frac{z^{-3}}{z^3} = \frac{3}{2} \times x^4 \times \frac{1}{y^4} \times \frac{1}{z^6} = \frac{3x^4}{2\,y^4z^6}$$

21) Answer: B

$$m = \frac{y_2 - y_1}{x_2 - x_1} = \frac{1-(-4)}{5-2} = \frac{5}{3}$$

$$y - y_1 = m(x - x_1) \rightarrow y - (-4) = \frac{5}{3}(x - 2)$$

$$y + 4 = \frac{5}{3}(x - 2) \rightarrow 3(y + 4) = 5(x - 2) \rightarrow 3y + 12 = 5x - 10$$

$$3y - 5x = -10 - 12 \rightarrow 3y - 5x = -22$$

22) Answer: C

$$\frac{x^2+7}{y-3} = \frac{(-5)^2+7}{7-3} = \frac{32}{4} = 8$$

23) Answer: C

Ethan paid $3.2 for two hours and $1.5 for each of three half-hour period after that.

$3 \times 1.5 = 4.5$

$3.2 + 4.5 = 7.7$

24) Answer: A

Cross multiply and isolate x: $\frac{c-d}{dx} = \frac{2}{7} \rightarrow 7(c - d) = 2dx \rightarrow x = \frac{7(c-d)}{2d}$

$$x = \frac{7}{2}\left(\frac{c}{d} - \frac{d}{d}\right) = \frac{7}{2}\left(\frac{c}{d} - 1\right)$$

25) Answer: D

$$9x^4 - 4x^2 = x^2(9x^2 - 4) = x^2(3x - 2)(3x + 2)$$

26) Answer: B

Factoring: $(x - 6)(x - 8) = 0 \rightarrow \begin{cases} x - 6 = 0 \rightarrow x = 6 \\ x - 8 = 0 \rightarrow x = 8 \end{cases}$

27) Answer: A

Two points are (0,4) and (10,6)

$m = \frac{y_2 - y_1}{x_2 - x_1} = \frac{6-4}{10-0} = \frac{2}{10} = \frac{1}{5}$

$y - y_1 = m(x - x_1) \rightarrow y - 4 = \frac{1}{5}(x - 0)$

$y - 4 = \frac{1}{5}x \rightarrow y = \frac{1}{5}x + 4$

28) Answer: D

$\begin{cases} 3 \times (-3x + 2y = -2) \\ 2 \times (4x - 3y = 5) \end{cases} \rightarrow \begin{cases} -9x + 6y = -6 \\ 8x - 6y = 10 \end{cases}$ →add two equations:

$-x = 4 \rightarrow x = -4$

29) Answer: A

Triangle third side rule: length of the one side of a triangle is less than the sum of the lengths of the other two sides and greater than the positive difference of the lengths of the other two sides.

the third side is less than 5+8=13 and greater than 8−5=3

30) Answer: D

Mean = $(111 + 120 + 127 + 140 + 159 + 165) \div 6 = 822 \div 6 = 137$

Only 148 can increase the mean.

31) Answer: C

The mode is the value which occurs with the greatest frequency. Oregon and New York are greatest and the same frequency (3 times).

32) Answer: B

convert the given distance, 8 km, into centimeters, (units on the map)

8 km = 8,000 m = 800,000 cm

divide by the ratio 1:50,000.

$\frac{800,000}{50,000} = 16 \text{cm}$

33) Answer: D

Area of square is: $a^2 = (\sqrt{13})^2 = 13$

THEA Math Practice Book

34) Answer: B

$2x^4y^4(3x^3y^3)^4 = 2x^4y^4(81x^{12}y^{12}) = 162x^{16}y^{16}$

35) Answer: C

Change percent to decimal: $25\% = 0.25$; $0.25 \times 220 = 55$

36) Answer: D

Using the formula for mean: Mean $= \dfrac{sum\ of\ the\ several\ given\ values}{number\ of\ value\ given}$

$= \dfrac{x+(x-3)+(x-3)}{3} = \dfrac{3x-6}{3} = \dfrac{3(x-2)}{3} = x - 2$

37) Answer: C

a value that is inversely proportional to another value: $a = \dfrac{k}{2b-5}$ (where k is a constant of proportionality)

substitute a and b: $8 = \dfrac{k}{2(9)-5} \rightarrow k = 8(13) = 104 \rightarrow a = \dfrac{104}{2b-5}$

38) Answer: B

Baseballs and basketballs are spherical. The volume of sphere: $v = \dfrac{4}{3}\pi r^3$

$36\pi = \dfrac{4}{3}\pi r^3 \rightarrow 36 = \dfrac{4}{3}r^3 \rightarrow 108 = 4r^3 \rightarrow r^3 = 27 \rightarrow r = 3$

$d = 2r \rightarrow d = 2 \times 3 = 6$

39) Answer: A

Use percent formula: Part $= \dfrac{percent \times whole}{100}$

$154 = \dfrac{percent \times 140}{100} \Rightarrow \dfrac{154}{1} = \dfrac{percent \times 140}{100}$, cross multiply.

$15,400 = percent \times 140$, divide both sides by 140. $\rightarrow$ percent$= 110$

40) Answer: D

The sample shows that 5 out of 600 phones will be faulty. Consequently, a proportion can be set up.

$\dfrac{5}{600} = \dfrac{P}{18,000}$ (P: Faulty Phone) $\rightarrow P = \dfrac{5 \times 18,000}{600} = 150$

41) Answer: B

We can Write an equation to solve the problem.

Emma Books = Mia books + 4 → Mia book = Emma − 4 = b − 4

Emma + Mia = 28

$b + b - 4 = 25 \rightarrow 2b - 4 = 28 \rightarrow 2b = 28 + 4 \rightarrow b = 16$

42) Answer: D

Probability = $\frac{number\ of\ desired\ outcomes}{number\ of\ total\ outcomes} = \frac{6}{11+6+7} = \frac{6}{24} = \frac{1}{4} = 0.25$

43) Answer: A

Surface Area of a cylinder = $2\pi r (r + h)$,

The radius of the cylinder is 4 inches, and its height are 9 inches. π is 3.14. Then:

Surface Area of a cylinder = 2 (3.14) (4) (4 + 9) = 326.56 inches

44) Answer: D

A line perpendicular to a line with slope m has a slope of $-\frac{1}{m}$.

So, the slope of the line perpendicular to the given line is $-\frac{1}{\frac{a}{b}} = -\frac{b}{a}$.

45) Answer: C

If 38% of the total number of customers is female, then 100% − 38% = 62% of the customers are male. Calculate 62% of the total. $0.62 \times 750 = 465$.

46) Answer: B

First, add the numbers given in the proportion to get a denominator: 8 + 3 = 11.

Then, the two parts can be represented as $\frac{8}{11}$ and $\frac{3}{11}$.

The smaller share is $\frac{3}{11}$ of 66.33 kg: $\frac{3}{11} \times 66.33\ kg = 18.09$

47) Answer: A

Since this is subtraction of 2 fractions with different denominators, their least common denominator is: $(x + 2)(4x - 1)$

$\frac{5}{(x+2)} - \frac{2}{(4x-1)} = \frac{5(4x-1)-2(x+2)}{(x+2)(4x-1)} = \frac{20x-5-2x-4}{(x+2)(4x-1)} = \frac{18x-9}{(x+2)(4x-1)}$

48) Answer: C

State the problem in a mathematical equation:

$2x = 6x - 20 \rightarrow 2x - 6x = -20 \rightarrow -4x = -20 \rightarrow x = \dfrac{-20}{-4} = 5$

$6x - 20 = 6(5) - 20 = 10$

49) Answer: C

$2h = 5p + 1$:

$p = 3 \rightarrow 2h = 5(3) + 1 = 16 \rightarrow 2h = 16 \rightarrow h = 8$

$p = 5 \rightarrow 2h = 5(5) + 1 = 26 \rightarrow 2h = 26 \rightarrow h = 13$

$p = 11 \rightarrow 2h = 5(11) + 1 = 56 \rightarrow 2h = 56 \rightarrow h = 28$

possible values of h: $\{8, 13, 28\}$

50) Answer: A

There are 6 digits in the repeating decimal (047619), so 4 would be the second, eighth, fourteenth digit and so on.

To find the 764th digit, divide 764 by 6.

$764 \div 6 = 127$ R2

Since the remainder is 2, that means the 764th digit is the same as the 2nd digit, which is 4.

THEA Practice Test 2
Answers and Explanations

1) Answer: C

The smallest prime number is 2, and the largest even negative integer is −2.

2+5 (−2) = 2−10 = −8.

2) Answer: B

The difference between his income and his cost is monthly saving.

January: $3,740− $2,202= $1,538

February: $3,855 − $1,985= $1,870

March: $3,970 − $2,005 = $1,965

3) Answer: D

State the problem in a mathematical sentence:

$a + 29 = 244 - 4a \rightarrow a + 4a = 244 - 29$

$5a = 215 \rightarrow a = 43$

4) Answer: A

$8\frac{7}{24} - 6\frac{2}{3} + 2\frac{5}{6} = (8 - 6 + 2)\frac{7}{24} - \frac{16}{24} + \frac{20}{24} = 4(\frac{-9}{24} + \frac{20}{24}) = 4(\frac{20-9}{24}) = 4\frac{11}{24}$

5) Answer: C

$3^3 + 3 + 3^0 = 27 + 3 + 1 = 31$

6) Answer: D

$\frac{153}{170} = \frac{153}{17} \times \frac{1}{10} = 9 \times \frac{1}{10} = 0.9$

7) Answer: A

Use the formula for Percent of Change:

$\frac{New\ Value - Old\ Value}{Old\ Value} \times 100\% = \frac{32.10 - 30}{30} \times 100\% = \frac{2.10}{30} \times 100\% = 7\%$

8) Answer: A

$|2x - 3| = 7 \rightarrow \begin{cases} 2x - 3 = 7 \rightarrow 2x = 10 \rightarrow x = 5 \\ 2x - 3 = -7 \rightarrow 2x = -4 \rightarrow x = -2 \end{cases}$

THEA Math Practice Book

9) Answer: C

Multiply equation (2) by 5. Add two equations [(1) +5(2)]:

$\begin{cases} 2x - 5y = 16 \\ 10x + 5y = 20 \end{cases} \to 12x = 36 \to x = 3$

Substitute $x = 3$ into equation (1): $2(3) - 5y = 16 \to -5y = 10 \to y = -2$

10) Answer: B

3 hours equals 180 minutes. The train passes 5 stations every 30 minutes: $\frac{5}{30}$

$\frac{5}{30} = \frac{x}{180} \to x = \frac{180 \times 5}{30} = 30$ stations

11) Answer: D

Difference in percent: $16\% - 11\% = 5\%$

$5\% \times 160 = 0.05 \times 160 = 8$

12) Answer: C.

$C = \pi d = 2\pi r = 2\pi \times 18 = 36\pi$

13) Answer: C

$V = l \times w \times h = 19 \times 15 \times 8 = 2{,}280\ in^3$

14) Answer: D

$\sqrt[4]{5^{-8}} = \sqrt[4]{\frac{1}{5^8}} = \frac{\sqrt[4]{1}}{\sqrt[4]{5^8}} = \frac{1}{5^{\left(\frac{8}{4}\right)}} = \frac{1}{5^2} = 5^{-2} = \frac{1}{25}$

15) Answer: A

There are no values of the variable that make the equation true.

16) Answer: C

Volume of cylinder: $V = \pi r^2 h = \pi \times 9^2 \times 8 = 648\pi$

Volume of cone: $V = \frac{1}{3}\pi r^2 h = \frac{1}{3}\pi \times 9^2 \times 6 = 162\pi$

$648\pi + 162\pi = 810\pi$.

17) Answer: D

$7.25 \times 10^{-5} = 0.0000725$

THEA Math Practice Book

18) Answer: C

$x + 52 + 42 + 36 = 180 \rightarrow x + 130 = 180 \rightarrow x = 50$

19) Answer: B

use the Pythagorean theorem to find the value of unknown side.

$a^2 + b^2 = c^2 \rightarrow 25^2 = a^2 + 15^2 \rightarrow a^2 = 625 - 225 = 400 \rightarrow a = 20$

20) Answer: C

Use percent formula: $\text{Part} = \frac{\text{percent} \times \text{whole}}{100}$

Part $= \frac{12 \times 150}{100} = 18$

Last price: $150 + 18 = \$168$

21) Answer: A

Area of trapezoid: $\frac{(a+b)}{2} \times h \rightarrow A = \frac{(12+14)}{2} \times 10 = 130$

Area of triangle: $\frac{b \times h}{2} \rightarrow A = \frac{7 \times 10}{2} = 35$

Area of shaded region: $130 - 35 = 95$

22) Answer: D

$xy - 8x = 54 \rightarrow x(y - 8) = 54$

$6x = 54 \rightarrow x = 9$

23) Answer: C

Multiply both sides by 4: $4 \times \left(\frac{x}{4}\right) = 4 \times (x - 6) \rightarrow x = 4x - 24$

Subtract both side by x and add 24 to both sides: $24 + x - x = 4x - x - 24 + 24 \rightarrow 3x = 24 \rightarrow x = 8$

24) Answer: D

$\begin{cases} 3 \times (-5x + 4y = 7) \\ -4 \times (-4x + 3y = 5) \end{cases} \rightarrow \begin{cases} -15x + 12y = 21 \\ 16x - 12y = -20 \end{cases}$ →add two equations:

$x = 21 - 20 \rightarrow x = 1$

THEA Math Practice Book

25) Answer: A

There are 3 parts labeled "F" out of a total of 8 equal parts.

The probability of not spinning at "F" is 5 out of 8.

26) Answer: C

The best estimate of the product of 32 and 5.04 to the nearest whole number.

Since 32 is already a whole number and 5.04 is closer to 5 than 6. Option C is the correct answer.

27) Answer: D

Fuel consumption rate $=\frac{7}{50}=0.14$ liter per mile

Cost: $1{,}500 \times 0.14 \times 1.40 = 294$

28) Answer: B

Point $1(x_A, y_A) = (-6, -9)$; Point $2(x_B, y_B) = (2, -3)$

Distance between two points = $\sqrt{(x_B - x_A)^2 + (y_B - y_A)^2}$

$\rightarrow d = \sqrt{(2-(-6))^2 + (-3-(-9))^2} = \sqrt{8^2 + 6^2} = \sqrt{64+36}$

$\rightarrow d = \sqrt{100} = 10$

29) Answer: A

Ordered values: 75, 89, 124, 198, 225, 280, 512.

The 4th number is median.

30) Answer: D

$\sqrt{4x-7} = 9 \rightarrow 4x - 7 = 81 \rightarrow 4x = 88 \rightarrow x = 22$

31) Answer: A

Get the equation into slop intercept form:

$y = mx + b$ where m is slope and b is y-intercept.

$x = 4y - 8$; Add 8 to both sides $\rightarrow x + 8 = 4y$; Dividing by 4: $\rightarrow y = \frac{1}{4}x + 2$

Thus, the y-intercept is 2.

THEA Math Practice Book

32) Answer: C

$3^{2x} = 729 \to (3^2)^x = 729 \to 9^x = 9^3 \to x = 3$

33) Answer: C

$18: 2 \times 3 \times 3$

$64: 2 \times 2 \times 2 \times 2 \times 2 \times 2$

LCM $(18, 64) = 2 \times 2 \times 2 \times 2 \times 2 \times 2 \times 3 \times 3 = 576$

34) Answer: B

Surface area $(SA): 6a^2 \to 96 = 6a^2 \to a^2 = 16 \to a = 4$ cm

Volume $(v): a^3 \to v = 4^3 = 64$ cm^3

35) Answer: D

Write the equation into standard form: $x^2 + 2x = 35 \to x^2 + 2x - 35 = 0$

Factor this expression: $(x - 5)(x + 7) = 0 \to x = 5$ or, $x = -7 \to x^2 = \begin{cases} 25 \\ 49 \end{cases}$

36) Answer: D

The minimum amount of water: $5 \times \frac{3}{5} = 3$; $3 \div 2 = 1.5$

The maximum amount of water: $5 \times \frac{3}{5} = 3$

Amount of water in all pitchers: $1.5 < w < 3$

37) Answer: C

The range is: highest value – lowest value.

R $= 65 - 17 = 48$

38) Answer: A

d $= 2r \to r = \frac{d}{2} = \frac{16}{2} = 8$

Area of circle (A): $\pi r^2 = \pi \times 8^2 = 200.96 \cong 201$

39) Answer: B

Factor the expression: $\frac{2x^2 - 15x + 7}{4(x^2 - \frac{1}{4})} = \frac{(x-7)(2x-1)}{4x^2 - 1} = \frac{(x-7)(2x-1)}{(2x-1)(2x+1)} = \frac{x-7}{2x+1}$

WWW.MathNotion.com

THEA Math Practice Book

40) Answer: D

$(\frac{1}{3})^{-4} = \frac{1}{3^{-4}} = 3^4 = 81$

41) Answer: B

$\frac{(5x^{-3}y^4)^3}{100y^{-2}z^2} = \frac{125x^{-9}y^{12}}{100y^{-2}z^2} = \frac{5y^{14}}{4x^9z^2}$

42) Answer: D

Rate: $\frac{\$1}{1Kg} = \frac{100¢}{1,000g} \to 1\,\$/kg = 0.1\,¢/g$

The price is: $30 \times 0.1 = 3$ cent per kilogram.

43) Answer: C

$3P = 6 \to P = 2$ The price of 2007 is one-third of 2012 ($6 \div 3 = 2$)

44) Answer: D

First equation: $4y + 5 = 20x + 9 \to 4y = 20x + 4 \to y = 5x + 1 \to m_1 = 5$

Second question: $5y + 4 = x + 7 \to 5y = x + 3 \to y = \frac{1}{5}x + \frac{3}{5} \to m_2 = \frac{1}{5}$

$m_1 = 5$ and $m_2 = \frac{1}{5}$, they aren't equal slopes or negative reciprocals.

45) Answer: C

Area = $5\frac{1}{3} \times 8\frac{2}{5} = \frac{16}{3} \times \frac{42}{5} = \frac{672}{15} = \frac{672}{15}$ square feet.

One-fourth = $\frac{672}{15} \div 4 = \frac{224}{5} \times \frac{1}{4} = \frac{224}{20} = \frac{112}{10} = 11.2$

46) Answer: A

Two points are $(2, 2)$ and $(-4, -1) \to m = \frac{y_2-y_1}{x_2-x_1} = \frac{-1-2}{-4-2} = \frac{-3}{-6} = \frac{1}{2}$

47) Answer: B

$V = \left(\frac{base \times height}{2}\right) \times$ height of prism $\to V = \left(\frac{10 \times 15}{2}\right) \times 5 = 375$

48) Answer: C

$f(a) = 4a - 7$ and $f(a) = -19$: $4a - 7 = -19$ so that $4a = -12$ and $a = -3$.

$f(b) = 4b - 7$ and $f(b) = 9$: $4b - 7 = 9$ so that $4b = 16$ and $b = 4$

THEA Math Practice Book

Finally, f(a + b) = f(−3 + 4) = f(1)

f(1) = 4(1) − 7 = −3.

49) Answer: D

The ratio of $x:y$ is equivalent to x divided by y, or $\frac{x}{y}$.

Equation: $28x = 12y$

Dividing both sides by 28: $x = \frac{12y}{28}$

Dividing both sides by y: $\frac{x}{y} = \frac{12}{28} = \frac{3}{7}$

Then, $x:y$ is $3:7$

50) Answer: B

The sum of exterior angles of any polygon is always equal to 360 degrees.

Exterior angle(degrees) $\times n = 360$; (n, is the number of angles (sides)).

Exterior angle = 180 − interior angle

The exterior angle = 180 − 168 = 12 degrees

Then, $12\ n = 360 \rightarrow n = 360 \div 12 = 30$

The polygon has 30 angles and 30 sides.

"END"

www.ingramcontent.com/pod-product-compliance
Lightning Source LLC
LaVergne TN
LVHW061308060426
835507LV00019B/2068